T0211516

SpringerBriefs in Applied Sciences and Technology

Series editor

Janusz Kacprzyk, Polish Academy of Sciences, Systems Research Institute, Warsaw, Poland

SpringerBriefs present concise summaries of cutting-edge research and practical applications across a wide spectrum of fields. Featuring compact volumes of 50–125 pages, the series covers a range of content from professional to academic.

Typical publications can be:

- A timely report of state-of-the art methods
- An introduction to or a manual for the application of mathematical or computer techniques
- A bridge between new research results, as published in journal articles
- A snapshot of a hot or emerging topic
- An in-depth case study
- A presentation of core concepts that students must understand in order to make independent contributions

SpringerBriefs are characterized by fast, global electronic dissemination, standard publishing contracts, standardized manuscript preparation and formatting guidelines, and expedited production schedules.

On the one hand, **SpringerBriefs in Applied Sciences and Technology** are devoted to the publication of fundamentals and applications within the different classical engineering disciplines as well as in interdisciplinary fields that recently emerged between these areas. On the other hand, as the boundary separating fundamental research and applied technology is more and more dissolving, this series is particularly open to trans-disciplinary topics between fundamental science and engineering.

Indexed by EI-Compendex and Springerlink.

More information about this series at http://www.springer.com/series/8884

Xu Guo · Gengdong Cheng
Wing-Kam Liu

Report of the Workshop Predictive Theoretical, Computational and Experimental Approaches for Additive Manufacturing (WAM 2016)

Springer

Xu Guo
Department of Engineering Mechanics
Dalian University of Technology
Dalian
China

Wing-Kam Liu
Department of Mechanical Engineering
Northwestern University
Evanston, IL
USA

Gengdong Cheng
Department of Engineering Mechanics
Dalian University of Technology
Dalian
China

ISSN 2191-530X   ISSN 2191-5318   (electronic)
SpringerBriefs in Applied Sciences and Technology
ISBN 978-3-319-63669-6   ISBN 978-3-319-63670-2   (eBook)
https://doi.org/10.1007/978-3-319-63670-2

Library of Congress Control Number: 2017953783

Printed on acid-free paper

This Springer imprint is published by Springer Nature
The registered company is Springer International Publishing AG
The registered company address is: Gewerbestrasse 11, 6330 Cham, Switzerland

Acknowledgements

We thank the National Natural Science Foundation of China (NSFC, Grant No. 11642016), the Chinese Society of Theoretical and Applied Mechanics (CSTAM) and the Innovation and Talent Recruitment Base on Numerical Simulation and Optimization of Rubber and Plastic Products Forming (B14013) very much for their sponsorship of this workshop. We would also like to thank the U.S. National Committee on Theoretical and Applied Mechanics (USNC/TAM) for its support of this workshop.

In addition, we wish to thank the following members for their contributions to this report: Professors Jun Yan, Zhao Zhang, Yongtao Lv, Hongfei Ye, Qinglin Duan, Yiqian He, Rui Li, Weisheng Zhang, and Yonggang Zheng at Dalian University of Technology.

Reviewers

We would like to thank the following individuals for their review of this report: Prof. Zhen Chen, University of Missouri; Prof. Zhuo Zhuang, Tsinghua University; and Prof. Zhanli Liu, Tsinghua University. We also would like to thank the invited speakers for their review of this report.

Although the reviewers listed above have provided many constructive comments and suggestions, they were not asked to endorse the content of the report, nor did they see the final draft before its release. Responsibility for the final content of this report rests entirely with the rapporteur and the institution.

Gengdong Cheng
Wing-Kam Liu
Xu Guo
Feng Lin
Planning Committee

International Research Center for Computational Mechanics, Dalian University of Technology

The **International Research Center for Computational Mechanics at Dalian University of Technology** (hereinafter referred to as the Center) is associated with the Department of Engineering Mechanics and the State Key Laboratory of Structural Analysis for Industrial Equipment at Dalian University of Technology. The center works at the international frontier in computational mechanics and the national major demand for scientific computation in engineering. The center provides a platform for international cooperation and academic exchange in computational mechanics in order to attract top-ranking scholars and train top-notch innovative talents in an international innovation base, thereby having great impact.

The center's objective is to promote institutional innovation and the reform of education/research systems in terms of building form, research organization, talent training, and performance evaluation. Through the center, the capability in original innovation and the academic reputation in international community will be strengthened by: (1) Building the strength of the discipline by supporting the development of the top-ranking discipline in computational mechanics and novel interdisciplinary research; (2) Improving the academic impact to reach the world first-class level by undertaking research projects with original innovation and significance in international frontier research or related to major demands in engineering computation; (3) Forming research groups with top academic leaders who have an international view and a commitment to outstanding innovation in computational mechanics; (4) Taking full advantage of international talent cultivation and further enhancing the quality of talent training; (5) Creating a world-class international facility in terms of operation, talent evaluation, academic assessment, and supporting services.

The Center has established several scientist studios. The research teams of the center consist of faculty members, guest scientists, and short-term domestic, and foreign visiting scholars.

Research Directions

- Computational methods for multiphysics, multiscale and coupled problems
- Optimization and design of advanced materials and structures
- Numerical simulation and design optimization of advanced manufacturing processes
- Analysis of design optimization under uncertainty
- Applications of computational mechanics in engineering and science

Planning Committee on Predictive Theoretical and Computational Approaches for Additive Manufacturing

GENGDONG CHENG (*Chair*), Dalian University of Technology
WING KAM LIU (*Co-Chair*), Northwestern University and Chair of the National Committee for Theoretical and Applied Mechanics within the U.S. National Academies of Sciences, Engineering and Medicine
XU GUO, Dalian University of Technology
FENG LIN, Tsinghua University

Contents

Chapter 1
Introduction

Additive manufacturing (AM) technique, as one of the leading forms of advanced manufacturing, is characterized by its flexibility to manipulate material compositions, structures and properties in end-use products with arbitrary shapes. AM has been successfully applied to conventional fields, e.g., biomedical devices and implants, aerospace components and rapid prototyping. At the same time, AM has the potential to promote transformative research in many fields across the vast spectrum of engineering and materials science. While experimental workshops in AM have been held previously, this workshop uniquely focused on theoretical and computational approaches and involved areas such as simulation-based engineering and science, integrated computational materials engineering, mechanics, materials science, manufacturing processes, and other specialized areas.

1.1 Workshop Overview

The Workshop on Predictive Theoretical, Computational and Experimental Approaches for Additive Manufacturing (WAM2016) was held October 17–19, 2016 in Dalian, China. This workshop was a continuation of "A Workshop on Predictive Theoretical and Computational Approaches for Additive Manufacturing" held October 7–9, 2015 in Washington, DC U.S. hosted by the USNC/TAM and sponsored by the National Science Foundation, the National Institute of Standards and Technology, and the Sandia National Laboratories. The WAM2016 was sponsored by the National Natural Science Foundation of China (NSFC), the Chinese Society of Theoretical and Applied Mechanics (CSTAM) and the Innovation and Talent Recruitment Base on Numerical Simulation and Optimization of Rubber and Plastic Products Forming. The workshop was designed to promote the international collaboration among researchers, educators and practicing engineers, and to address the challenges and opportunities in theoretical,

© The Author(s) 2018
X. Guo et al., *Report of the Workshop Predictive Theoretical, Computational and Experimental Approaches for Additive Manufacturing (WAM 2016)*, SpringerBriefs in Applied Sciences and Technology, https://doi.org/10.1007/978-3-319-63670-2_1

computational and experimental methods to advance additive manufacturing in a holistic, multifaceted and interdisciplinary way.

The workshop provided a platform to exchange information regarding recent advances related to AM; promote international collaborations in research and education; and make an immediate contribution to the progress of additive manufacturing by disseminating new findings and integrating different team efforts among active researchers, educators and practicing engineers.

Prof. Dongming Guo, the President of Dalian University of Technology (DUT), opened the workshop with a brief introduction of DUT. He hoped that the workshop can promote the international collaboration between DUT and governments, universities, research institutes and industries.

Prof. Wei Yang, the President of NSFC and CSTAM, announced that NSFC and CSTAM fully support activities and additive manufacturing collaborations between the U.S. and China, as well as the collaborations between CSTAM and U.S. National Committee in TAM. According to Prof. Yang, AM currently takes the center stage of manufacturing sciences and it undergoes a transition period from "NOVEL" (Numerous in applications, Organic in the first generation of base materials, Versatile in functions and Economical and Lightweight in cost and material saving) to "GREAT" (Geometric supremacy, Rapid in design and manufacturing, Endurance in harsh service environment, Accuracy of additive manufacturing and Transformable in shape).

Prof. Gengdong Cheng, chairman of the workshop, introduced the research activities of computational mechanics in DUT, which was initiated by Prof. Lingxi Qian, by three examples, i.e., the change of the design of Three Gorges Shiplift, the light weight design of rockets and the concurrent multiscale topology optimization of structure and material.

Prof. Wing-Kam Liu, co-chairman of the workshop, gave a detail explanation on the background, task and organization of the workshop and described the topics of the subgroup discussions.

1.2   Organization of This Report

Subsequent chapters of this report summarize the workshop presentations and discussions. Chapter 2 provides an overview of the theoretical understanding of material science and mechanics in AM. Chapter 3 focuses on computational and analytical methods in AM. Chapter 4 discusses theories, methods and tools for AM oriented design and optimization. Chapter 5 focuses on AM experimental methods and results, and AM scalability. Chapter 6 describes discussions regarding some fundamental issues that need to be addressed in the future work. Finally, the workshop speakers are listed in Appendix A and the workshop agenda is presented in Appendix B.

Chapter 2
Theoretical Understanding of Material Science and Mechanics in Additive Manufacturing

The first session of the two-day workshop provided an overview of the workshop and additive manufacturing (AM), and some theoretical and simulation studies on AM. These included the appealing advantages and features of AM for products development, developmental trends in AM, and the multiscale multiphysics modelling framework of AM. Dr. Kathie Bailey (National Academies of Sciences, Engineering and Medicine), Prof. Bingheng Lu and Prof. Ting Wen (Xi'an Jiaotong University), Prof. Wing Kam Liu (Northwestern University) respectively, discussed research results, challenges and future directions related to the following questions:

- What are the current research and driving forces on AM in U.S. and China?
- Why multiscale modeling is important in AM?
- Which are the available/best methodologies to bridge the spatial and temporal scales in AM?
- How do we best keep track of the material states?
- Which is the more suitable mechanistic constitutive model?
- Why defect predictions are an important issue in AM?
- What are the most important physical features to capture in AM?

2.1 Overview of the U.S. Workshop on Predictive Theoretical and Computational Approaches for Additive Manufacturing

In this talk, Dr. Kathie Bailey provided an overview of the U.S. workshop on predictive theoretical and computational approaches for additive manufacturing. She emphasized that international collaborations are important to the United States.

© The Author(s) 2018
X. Guo et al., *Report of the Workshop Predictive Theoretical, Computational and Experimental Approaches for Additive Manufacturing (WAM 2016)*, SpringerBriefs in Applied Sciences and Technology, https://doi.org/10.1007/978-3-319-63670-2_2

For example, the National Academies of Sciences, Engineering, and Medicine serves as the U.S. adhering body to a number of international scientific organizations, including the International Council for Science (ICSU) and many of its member scientific unions, such as the International Union for Theoretical and Applied Mechanics (IUTAM). She pointed out that the National Academies carries out their responsibilities to these organizations through a set of related U.S. National Committees (USNCs) which serve as a conduit of the information between the U.S. disciplinary community and the international unions. The USNCs also organize workshops which are of interest to their respective U.S. and international communities.

Dr. Bailey pointed out that the U.S. Workshop on Predictive Theoretical and Computational Approaches for Additive Manufacturing was organized by the USNC for Theoretical and Applied Mechanics and held on October 7–9, 2015 in Washington, DC. The workshop was sponsored by the National Science Foundation (NSF), Sandia National Laboratories, and the National Institute of Standards and Technology (NIST), and the workshop was focused on four topics: theoretical understanding of materials science and mechanics; computational and analytical methods in AM; monitoring and advanced diagnosis to enable AM fundamental understanding; and scalability, implementation, readiness, and transitions. The U.S. workshop was attended by 50 persons and webcasted to nearly 200 additional participants. The related videos received over 1700 views and are available on the workshop website. The U.S. workshop report was published and is on BISO website (www.nas.edu/biso). Printed copies are also available upon request.

2.2 Additive Manufacturing and Design Innovation

Prof. Bingheng Lu's talk was focused on 'additive manufacturing and design innovation'. After a brief introduction of AM, also known as 3D printing, Prof. Lu discussed how AM improves product development and expands design freedom and creativity. He concluded with a discussion on development trends in AM.

Prof. Lu noted that AM appeared around 30 years ago, and has caused a paradigm shift in manufacturing, as compared to traditional equal-material or reductive manufacturing. AM was ranked 9th of 12 disruptive technologies reported by McKinsey Global Institute in 2013, with potentially 0.2–0.6 trillion U.S. dollars economic impact in 2025.

Prof. Lu presented that AM enables fully integrated product and process design and development, where it only takes 1/3 to 1/10 of the time and cost of conventional development. A 3D printed auto part, used in Ford's 3.5L EcoBoost race engine, was shown as an example to demonstrate how AM facilitates the improved quality and optimized weight for better fuel efficiency, with months of development time and millions of dollars saved. Various aspects of how AM improved product development were discussed, i.e., design evaluation, assembling evaluation, aerodynamic verification, digital modeling and structure verification, democratization of design, and internet plus 3D printing. As a meaning of rapid prototyping, AM has also been used for design evaluation, where early visibility and early detection of deficiencies can be enhanced, while design and development cycles are compressed. Prof. Lu presented a dynamic 3D digital speckle strain measurement and analysis system, where 3D full field strain measurement can be used to identify strength bottleneck part of the product and thus improve the design. At the end of this part, he presented a future perspective of the opportunities for mass entrepreneurship and innovation through internet plus 3D printing, with new business models for customized manufacturing.

Prof. Lu also noted that AM dramatically expands the design freedom and creativity with various advantageous features, i.e., producing objects that is impossible by conventional manufacturing, increasing the complexity of objects without increasing production costs, and allowing the design and structural optimization for performance and the integration of multiple parts into one single part. A few examples were given to demonstrate these features, including: a 3D printed electric car named Urbee II and the Airbus concept plane and cabin to show integrated manufacturing with better and lighter design; 3D printed blade and jet engine fuel nozzle with hollow structures to show how 3D printing allows non-conventional variable shapes optimized to increase the heat exchanger's efficiency with extreme internal complexity; a 3D printed composite to show multi-material manufacturing with controlled shape and functionality; manufacturing of functionally graded structures optimized with less weight but better performance without extra cost.

Then Prof. Lu discussed a variety of topics that will shape the development of trends in AM (see Fig. 2.1), including

- Composite 3D printing, where understandings of the stress-strain relationship of the non-equilibrium interface, local interface of multi material, multi scale functional gradient material are required;
- Micro-nano scale 3D printing, i.e., 3D printing of composite structures across multiple length scales from nano to micro to macro with precision;
- 4D printing, where printed objects reshape themselves or self-assemble over time post printing, facilitated by heating, light, swelling in a liquid, etc.
- 3D bioprinting, i.e., 3D printing of biocompatible materials and live cells into complex 3D functional living tissues for tissue regeneration.

An overview of the development trends in AM was presented, covering different aspects of AM (technology, application, material, industry, participants, etc.) and a collaborative innovation was recommended to move AM forward. Prof. Lu commented that the future of AM will be limitless with 'any material' (plastics, ceramics, alloy, etc.), in 'any shape' (hollow parts, etc.), applicable to 'any field' (aerospace, biomedicine, automobiles, etc.), at 'any workplace' (office, home, etc.) and in 'any quantity' (personalized product, mass production), so called 5 'any's.

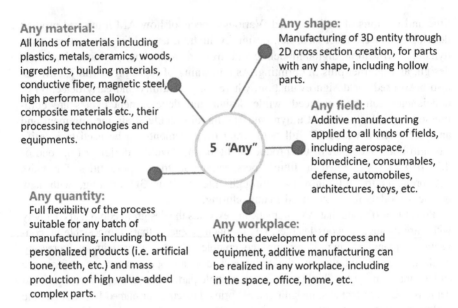

Any material:
All kinds of materials including plastics, metals, ceramics, woods, ingredients, building materials, conductive fiber, magnetic steel, high performance alloy, composite materials etc., their processing technologies and equipments.

Any shape:
Manufacturing of 3D entity through 2D cross section creation, for parts with any shape, including hollow parts.

Any field:
Additive manufacturing applied to all kinds of fields, including aerospace, biomedicine, consumables, defense, automobiles, architectures, toys, etc.

5 "Any"

Any quantity:
Full flexibility of the process suitable for any batch of manufacturing, including both personalized products (i.e. artificial bone, teeth, etc.) and mass production of high value-added complex parts.

Any workplace:
With the development of process and equipment, additive manufacturing can be realized in any workplace, including in the space, office, home, etc.

Fig. 2.1 Development trends in additive manufacturing. *Source* Prof. Bingheng Lu, Xi'an Jiaotong University, presentation to the workshop

A transition is already seen in manufacturing industry from tripartite confrontation of three manufacturing technologies (equal-material, reductive, additive) to three key parts with shared value under one roof.

In closing, Prof. Lu reminded workshop participants that:

- Additive manufacturing, as a meaning of rapid prototyping, enables quick evaluation and implementation of new design idea.
- Additive manufacturing opens up the capabilities for customized products in a wide variety of applications.
- Additive manufacturing eliminates constraints associated with conventional manufacturing, enabling us to re-examine the existing products and equipment design, to re-design from the point view of additive manufacturing, hence opening up a huge space for innovative design.

2.3 Deriving Process-Structure-Properties Relationships for Additive Manufacturing with Data-Driven Multi-Scale Multi-Physics Material Models

Prof. Wing-Kam Liu pointed out that AM possesses appealing potentials for locally or globally manipulating material compositions, structures and properties in end-use products with arbitrary shapes without the need for specialized tooling.

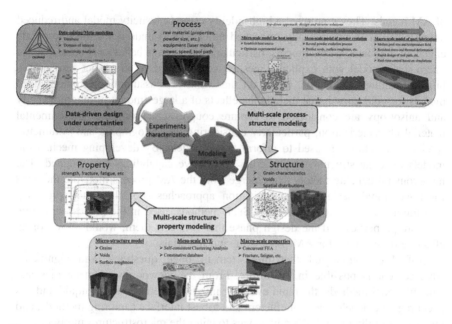

Fig. 2.2 Illustration on deriving process-structure-properties relationships for additive manufacturing with data-driven multi-scale multi-physics material model

Prof. Liu explained that since the multiple physical processes at multiple temporal and spatial scales are difficult to experimentally measure, numerical modeling is a powerful tool to shed light on the underlying physical mechanisms. His talk presented the latest work using data-mining techniques to close the cycle of design-prediction-optimization, based on comprehensive material modeling of process-structure-property relationships for AM materials (see Fig. 2.2).

Prof. Liu and his research team considered process-structure and structure-properties models. The multiscale process modeling started from micro-scale models of the electron- or photon-material interactions used to generate heat. Next, they modelled the meso-scale, consisting of individual powder particle evolution driven by fluid flow and thermal diffusion. The meso-scale model incorporated surface tension force, the Marangoni effect due to the temperature gradient, evaporation and recoil pressure, as well as a material-dependent and experiment-specific heat source model derived from the micro-scale model. Finally, for the macro-scale model, they used a specialized finite element analysis code with an outstanding computing efficiency to resolve the fabrication process of a complex product. They incorporated the meso-scale simulation results to enrich the predictability of the simplified macro-scale model. The model can also include concurrently coupled models at critical locations to capture more detailed information, such as fcc/bcc phase fraction.

To link structure and properties, Prof. Liu and his team used mechanistic models at several scales to capture a variety of behaviors. These models include factors such as voids, inclusions, and grain structures, which are the differentiating features of AM metals. They employed a high-fidelity crystal plasticity model and reduced order model with dramatically improved efficiency to understand the microscale effects of the material constituents, particularly on damage behavior. For instance, a micro-scale model with explicitly meshed voids was used to predict zero probability of failure fatigue life, based on the worst-case assumption. Homogenized models for void growth that include the effects of a large amount of inhomogeneity and anisotropy are considered to capture coupon-scale response. Experimental material characterization, particularly considering crystallographic and volumetric spatial distributions, is used to inform these models. By developing mechanistic models to relate structure to properties, predictive capabilities are obtained. The numerous influencing factors that emerge from the AM process motivate the need for novel rapid design and optimization approaches. For this, they proposed data-mining as an effective solution. Such methods, used in the process-structure, structure-properties and the design phase that connects them, would allow for an effective design loop for AM processing and materials.

Prof. Liu commented that there remain open questions, and significant advancement is possible in the future. From the viewpoints of mechanics and computational methods, the rapid evolving interfaces between solid, liquid and gas encourage the development of efficient and robust interface capturing methods, and the desire to fully exert AM's advantages to tailor the microstructures motivates the grain growth models with a much higher accuracy. AM is believed to open new avenues to materials design and structure optimization. A few exciting directions identified by Prof. Liu are functional graded materials, multi-material composites, and meta-materials. He emphasized that nano-scale AM together with nano-CALPHAD is also worthy of research efforts. He stated that these new directions open opportunities for AM in a wide range of sectors, from biomedical to automotive to aerospace.

2.4   Discussion

Following the panel's presentation, Dr. Kathie Bailey, Prof. Bingheng Lu and Prof. Wing Kam Liu participated in a discussion session moderated by the session chair, Prof. Xu Guo from Dalian University of Technology.

Prof. Lu was asked how to educate the young generation on AM, given that it is a newly emerging technique. Prof. Lu explained that AM by nature itself is an intuitive technology and thus can be easily understood by public. The low cost of 3D printers enables their accessibility. Because they are fun to use, the younger generation is attracted. Prof. Lu further added that education can be started from very young age (e.g., school kids) to adults (e.g., college students). Because of the wide variety of applications, he believes that it is crucial to attract talents with different backgrounds and academic disciplines to be involved in the AM technology. For instance, the 3D printing of human tissue requires collaborative work between engineers, medical doctors, and biologist.

Another question was raised regarding the 5 'any's. The concern was that the 5 'any's could lead to many problems such as illegalization or intellectual property violations. How can these problems be solved in the future? Prof. Lu commented that we need to work together with the companies to standardize the whole procedure; and meanwhile, we need to work with the local and national governmental agencies on beneficial policies.

Prof. Wing-Kam Liu was asked about the numerical simulation algorithm for AM on the continuum level. Prof. Liu commented that many numerical methods have been developed over the past 30 years, and many commercial software (e.g. FLOW3D) have been developed based on these codes. In the past 5 years, people converted the codes for the applications in AM, and a lot of new developments have been made, and consequently commercial codes became available. He further added that his research team includes skillful people who are responsible for developing the code for solving equations in AM.

The next question was regarding manufacturing defects, and how dangerous these defects are, and how they can be avoided. Prof. Liu explained that defects are an unavoidable problem, even for manufacturing components with simple geometry. He said that we may be able to manufacture components with no defect for a small part, but at the present time, it is not possible for the whole structure. He then used an example to explain the AM process and mentioned that at the present time, nobody can claim they are able to develop a process that is universal. Prof. Liu and his research team have spent years working to predict the products for one material, one configuration. but if we change the material, then we need to spend another 3–4 years to optimize the process. He further added that computational simulation is very important for the understanding of defects and its fundamental problems in AM. Numerical simulations can provide information which cannot be obtained from the physical experiments.

Another question raised is how to improve the performance of product (e.g. the titanium alloy for turbine) through AM, compared to the traditional manufacturing method, e.g. metal forming and forging. Prof. Liu commented that melting, metal forming and forging have existed for years. He said people believe that by melting and re-melting grains, refinement can be achieved to give stronger materials. Thus he believes that good performance of products may be achieved through AM in the near future.

Another participant raised a question regarding the existing commercial software available in the market and wondered which software is the best for AM. Prof. Liu commented that there are many companies selling software for the applications in AM. Because every software has its own advantages, the decision must depend on the task to be performed. He also mentioned that just like the development of other well-developed commercial software, there could be hypes in the current AM software because it is at the early development stage. He finished his talk by using an example to illustrate that the performance of software really depends on what people want to do, and for some tasks, no software is available at the moment.

Chapter 3
Computational and Analytical Methods in Additive Manufacturing

The second to fourth sessions of the first two days of the workshop were focused on the computational and analytical methods in additive manufacturing (AM). Prof. Ernst Rank (Technische Universität München), Prof. Weidong Huang (Northwestern Polytechnical University), Prof. Feng Lin (Tsinghua University), Prof. Chunze Yan (Huazhong University of Science and Technology), Dr. Alonso Peralta (Honeywell Aerospace Inc.), Prof. Huiling Duan (Peking University), Prof. Michele Chiumenti (Technical University of Catalonia), Prof. Lars-Erik Lindgren (Luleå University of Technology), Prof. Zhao Zhang (Dalian University of Technology) respectively, discussed research results, challenges and future directions related to this topic.

3.1 Simulation for Additive Manufacturing: An Adaptive Multi-scale and Multi-physics Approach

Prof. Ernst Rank began his presentation by introducing the advantages and challenges of Selective Laser Melting (SLM). He emphasized that SLM is extremely promising, since it allows: Large design freedom, complex geometry, customized parts and material design.

The physical phenomena during the SLM process is especially complex. First, the powder roller spreads metal powder layer (30–100 mm) on the base plate, and the laser beam fully melts the powder, then, fused material solidifies during cooling. Many physical phenomena occur, for example: phase change, vaporization, convection, volume shrinkage, fluid flow, radiation, conduction, Marangoni effect, melting/solidification, Prof. Rank explained. Thus, great challenges grow out of this, e.g.:

© The Author(s) 2018
X. Guo et al., *Report of the Workshop Predictive Theoretical, Computational and Experimental Approaches for Additive Manufacturing (WAM 2016)*, SpringerBriefs in Applied Sciences and Technology, https://doi.org/10.1007/978-3-319-63670-2_3

(1) Residual stress
(2) Dimensional accuracy of the product.

To predict the process and do the optimization, it is significant to simulate temperature evolution and residual stress of SLM additive manufacturing by numerical computation. There are many challenges in simulation of SLM (product scale), for example:

Multi-physics: Melting & solidification, thermo-mechanical problem and plasticity.

Evolving domain: Phase change and new powder layers.

Spatial scales: Laser radius ~ 0.1 mm and size of product $\sim 10 \times 10 \times 10$ cm^3.

Temporal scales: Phase change $\sim$ ms and complete process $\sim$ hours.

Several challenges have already been solved. One can use a temperature-based phase change model in which the heat equation is expressed in terms of enthalpy (H) and temperature (T) to deal with the phase change. Prof. Rank commented that one of the greatest challenges in simulation of additive manufacturing is the extreme multi-scale nature of products and processes. Considering e.g. the SLM, relevant spatial scales range from sub-micrometers (the width of the phase front of the melting pool) to kilometers (the length of the path of a laser to produce an artefact). Temporal scales range from microseconds to hours. Being aware of the multi-physics nature of the process it is obvious that a full resolution of all spatial and in particular all temporal scales will be out of reach even for Exascale Computing facilities, Prof. Rank explained, which can be expected within the next ten or twenty years.

Then he discussed principles of a split of scales and concentrate on the meso-scale, considering the evolution of temperature fields being coupled via a phase transition model to a thermo-elasto-plastic model. The goal is to compute the evolution of the solid domain as well as residual stresses and corresponding deformations. Considering the local problem in the vicinity of the laser only, many commercial finite element packages are available providing a wealth of models to simulate this multi-physics problem. State-of-the-art is the 'quiet element method' (see Michaleris 2014) which activates a previously already defined, yet inactive element for computation, as soon as the laser path is traversing it and changes the physical properties from a value mimicking a void space to a physically meaningful value of the generated solid. A scale analysis yet shows that for a typical SLM process the size of an element discretizing the coupled local problem with sufficient accuracy should be in the order of 10^4 times the diameter of the final specimen. It is obvious that a uniform finite element mesh resolving these scales is by far out of reach of any transient computation.

As a consequence of this drastic mismatch between capabilities of available commercial software and needs for an accurate computation some recent approaches suggest an octree-based mesh structure allowing for a dynamic refinement and de-refinement of a finite element mesh with low order elements. Prof. Rank took in his approach one step more, and applies hp-refinement (Zander et al. 2015), allowing at the same time for a local adaptation of the mesh (h-refinement) and of the shape functions' polynomial degree (p-extension). This combination of refinement and de-refinement has turned out to provide a very efficient and highly accurate discretization scheme suited in particular for problems including a large range of spatial scales. Most important is the observation that the number of degrees of freedom remains nearly constant over time and does not increase with the evolution of the printed structure. Furthermore he extend this approach to the so-called Finite Cell Method (FCM) (Kollmannsberger et al. 2016), which is a fictitious domain approach and is particularly well suited for problems of evolving domains, if using the p-version of the Finite Element Method as its basis. It allows for a spatial resolution of the solidified material independent of the computational mesh on an octree-like material grid being up to one order finer than the size of an element. With this decoupling of material description and approximation spaces for the field equations they gain significantly more freedom in discretizing the multi-scale process, compared to more classical low order methods (Fig. 3.1).

Yet, even an extremely powerful multi-scale performance in space is not sufficient to simulate the coupled process over the full range of temporal scales, according to Prof. Rank. Reduced order models or sparse grid methods (Butnaru 2013) are good candidates for closing this gap to a truly predictive simulation of SLM-processes, yet still need to show their feasibility in the future.

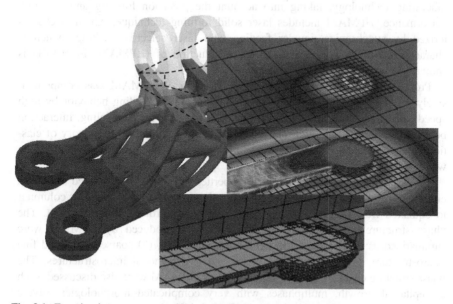

Fig. 3.1 Zooming into a structure with independently refined grids for thermal and elastoplastic fields and material description

Prof. Rank raised some major research challenges we have to deal with at present:

(1) Upscaling in space and time;
(2) Combine (physical) simulation and Data Analytics;
(3) CAD to part integration;
(4) Experimental validation;
(5) First time right design.

At last, Prof. Rank introduced Heinz Maier-Leibnitz neutrons source in Garching near München, where the First ECCOMAS International Conference on Simulation for Additive Manufacturing will be held in 2017, organized by E. Rank, F. Auricchio and P. Steinmann.

3.2 High Performance Metal Additive Manufacturing and Its Numerical Modelling

Prof. Weidong Huang provided an overview of High Performance Metal Additive Manufacturing (HPMAM) in Northwestern Polytechnical University, China (NWPU) (Huang and Lin 2014; Wei et al. 2015). HPMAM is an integrated manufacturing technology, taking into account the precision forming and the high performance. HPMAM includes laser solid forming technique, which is characterized by synchronize powder feeding, and laser selective melting, which is characterized by powder bed. A brief introduction of (HPMAM) in NWPU is shown in Fig. 3.2.

Prof. Huang reviewed that the precise forming of HPMAM was comprehensively investigated through characterizations of powder feeding behavior by high speed camera measurements and gas-powder two phase flow modeling. Interaction between powder particle and melt pool was revealed based on the theory of elasticity and force analysis, and the critical condition for adhering powder formation was obtained.

He explained that high mechanical properties of HPMAM are related to defects elimination, solidification processes optimization in the melt pool, such as columnar to equiaxed transition (CET) and microstructure control by heat treatment. The phase structure characteristics of titanium alloys produced by HPMAM were summarized as: (1) no dendrites (2) Al supersaturated, (3) coarse grains and fine microstructure, (4) new interface phase and (5) tri-modal microstructures. The phase structure characteristics of superalloys by HPMAM were also discussed, such as: epitaxial growth, multiphases with very complicated morphologies, severe micro segregation and easy to form cracks. The solid phase transformation of steels by HPMAM is significant influenced by thermal history. Based on the researches of these alloys phase structures, high mechanical property products by HPMAM were fabricated for industry applications, as can be seen in Fig. 3.2.

HPMAM in NWPU since 1995

- **Focusing on MAM of metallic materials**
- **Chasing very high mechanical properties with precise geometry**
- **Fousing on application in aviation and space industries**
- **Laser rapid forming (LRF) ⸺ Laser solid forming (LSF)**
- **SLM technology has been newly developed**
- **Brighter Laser Technology Co. Ltd (BLT) based on HPMAM technology of NWPU ⸺ focusing on industrial application of HPMAM**

Fig. 3.2 Brief introduction of high performance metal additive manufacturing (HPMAM) in Northwestern Polytechnical University, China (NWPU)

Prof. Huang displayed that a quantitative cellular automaton (CA) model was established for the multiscale numerical simulation of solidification in the melt pool, as shown in Fig. 3.3. The mesh induced artificial anisotropy of CA model was quantitatively estimated. The average artificial anisotropy in CA model is 0.56%, which is a little larger than that of phase field (PF) model (0.3%). However, the present CA model is 6 times faster than the PF model. The quantitative CA model can reproduce competitive bi-crystal directional solidification, lamellar and anomalous eutectic growth.

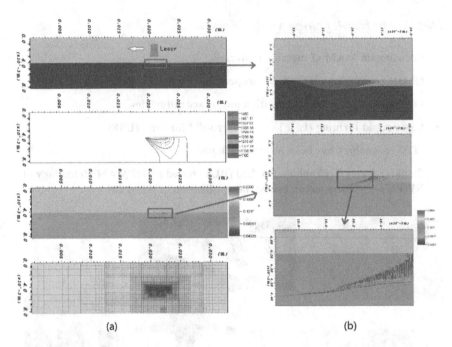

(a)   (b)

Fig. 3.3 Multiscale numerical simulation of solidification in the melt pool: (**a**) Macroscopic view of simulations results, including separations of phases, thermal/solute field and meshes; (**b**) Mesoscopic view of simulation results, including melt pool shape and solidification microstructure

3.3   Current Studies on Powder Bed Based EBSM Process

* Feng Lin recognized contributions from the following researchers: W. T. Yan, Y. Qian, and B. Zhou.

Prof. Feng Lin provided a self-developed electron beam selective melting (EBSM) system with the dual metal powder delivery capability, by which a Ti6Al4V-Ti47Al2Cr2Nb gradient material structure was made. EBSM is a power bed based additive manufacturing technique started in 2004 and is similar as EBM from Arcam AB, which features its higher heating efficiency, higher temperature in building chamber, lower thermal residual stress and the potential to adapt more material than SLM. New material was synthesized by melting the mixture of these two powders with fine and crack free microstructure. The particle based model was used to simulate and understand the EB scan process. Additionally, according to their studies on EBSM process simulation, he proposed the concept of "efficient energy" to synthesize variant process parameters, which will deal with both the energy input by EB scan and the energy dissipation through thermal conductivity, radiation or evaporation, to indict the necessary energy for the particle melting and deposition.

Firstly, Prof. Lin introduced the history of Tsinghua University's AM group simply, which was found by Professor Yongnian Yan in 1991. Now, the AM group mainly includes two centers with different research areas. The first one is the Bio-manufacturing center directed by Professor Wei Sun, who is a professor of National "Thousand Talent Plan". The second one is the Rapid forming center directed by himself. Next, he presented the fundamental dispersion-accumulation principle of the AM as shown in Fig. 3.4, which illustrated that, in the dispersion process, the 3D digital model is dispersed into layers, line segments and spots etc. degraded digital objects, then in the accumulation process, the degraded digital objects are transfer to physical objects and stacked to form a 3D physical model. From this principle, the additive manufacturing possesses several unique features and could be regarded as digital manufacturing which can achieve personalized and remote manufacturing, degradation manufacturing which can achieve the high capability for complex and monolithic structures manufacturing directly, accumu-lative manufacturing which brings advantages to the heterogeneous or multi-material manufacturing, integrative manufacturing which combines the material synthesize and material forming to solve for manufacturing difficulties of high performance materials or hard to process material, and rapid manufacturing which benefits the quick product development and in situ manufacturing.

Prof. Lin then reviewed the development on EBSM system. He said that the research on EBSM technique was started in 2004, which used EB, instead of laser, to scan the metal powder bed. Comparing with the laser, the EB owns following advantages:

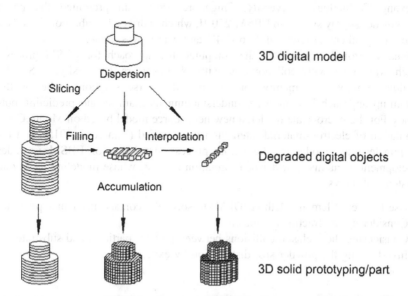

Fig. 3.4 Schematic diagram of the accumulation principle for 3D additive manufacturing. *Source* Prof. Feng Lin, Tsinghua University, presentation to the workshop

- The EB has higher efficiency and intensity than laser.
- The EB has higher and constant energy absorption of variant material.
- The EB is suitable for wide material.
- The EB could provide high powder bed temperature to lower the thermal stress.

Several generations of EBSM system have been developed in the past 12 years. The primary EBSM-150 was developed in 2004. In 2009, the second system, named as EBSM-250, was developed for Northwest Institute for Nonferrous Metal Research (NIN) and possessed the building size of 200 mm × 200 mm × 200 mm, in which TC4 parts and TiAl based alloy blade were fabricated by NIN. The third system, EBSM-250-II, was developed for the Hefei Institute of Physical Science (HIPS) of Chinese Academy of Sciences (CAS) in 2011, which is the first powder bed based additive manufacturing system with the dual-metal supply and gradient structure forming capacity.

Comparing with the SLM, the EBSM has some disadvantages. For instance, the beam spot of EB is about 200 μm which is bigger than the laser beam spot in SLM about 80 μm, and the surface roughness of the part made by EBSM is much higher than by SLM (Rafi et al. 2013).

Prof. Lin said that the gradient structures like the copper-steel gradient structure or TC4-TiAl gradient structure would have widely potential applications. He summarized the researches on powder bed based additive manufacturing process for the gradient structure. For example, the powder mixing subsystem for SLS was developed by the University of Texas at Austin. The Z-graded structures built by SLM was developed by the Universidad Federal de Santa Catarina, Brazil. The multi-materials processing by SLM (316 l/c18400 copper) was developed by Nanyang Technology University, Singapore. Prof. Lin presented the novel dual-powder supply system in EBSM-250-II, which is driven by vibration, based on which the gradient structure of Ti-6Al-4V and γ-TiAl was made.

Finally, Prof. Lin presented the computational approach for EBSM process, which was done with collaboration with Northwestern University, U.S. The top-down approach is mainly for design and inverse solutions, whereas the bottom-up approach is mainly for understanding mechanism and predicting outcomes. For the micro-scale model, a new heat source model based on Monto Carlo Simulation of electron-material interaction was applied (Yan et al. 2015), and for the powder-scale modeling, the whole process simulation of EBSM is under development, including powder bed formation process, whose modeling technique is listed as follows:

- Use Discrete Element Method (DEM) to solve the contact mechanics problem.
- Considering the fractional contacts.
- Considering the inelastic collisions between powder particles and substrate.
- Investigating the powder size distribution by experiment.

and the multi-physics modeling of powder evolution processes are listed as follows:

- Using Finite Volume Method to solve fluid flow equations.
- Using Volume of Fluid (VOF) to capture free surfaces.
- Considering the evaporation and the corresponding recoil pressure.
- Considering the Marangoni effect.
- Considering the buoyancy flow.

Prof. Lin showed some simulation results of the influence of main fabrication parameters, like EB power, scanning speed, and size of powder, etc. One of his simulation indicated that the smaller powder particles usually lead to better surface roughness.

At last, he proposed the concept of efficient energy, which could synthesize variant process parameters, based on the energy input and loss, to indicate the necessary energy for raising the particle temperature, melting the particles, and avoiding the defects.

3.4 Advanced Lightweight Metal Cellular Lattice Structures Fabricated via Selective Laser Melting

[*] Chunze Yan recognized contributions from the following researchers: Liang Hao and Yusheng Shi.

Prof. Chunze Yan reviewed that SLM is an AM process, which can directly make complex three-dimensional metal parts according to a computer-aided design (CAD) data by selectively melting successive layers of metal powders. SLM has the capability of producing structures of complex freeform geometry. It has been demonstrated to manufacture cellular lattice structures with fine features, showing a great potential to make advanced lightweight structures and products which are highly desired by engineering sectors such as aerospace, automotive and medical industries (Yan et al. 2015). However, SLM requires support structure to build overhang section if its angle from the horizontal is less than certain degree. This introduces design and manufacturing complications for the SLM of lightweight cellular structures and engineering components (Fig. 3.5).

He emphasized that the triply periodic minimal surfaces (TPMS) are smooth infinite surfaces that partition the space into two labyrinths in the absence of self-intersections, and are periodic in three independent directions. TPMS are ideal to describe the aforementioned biomorphic geometry. In their works, Gyroid and Diamond TPMS lattice structures were manufactured by SLM, and the manufacturability, microstructure and mechanical properties of the Ti-6Al-4V TPMS lattices were investigated (Yan et al. 2015).

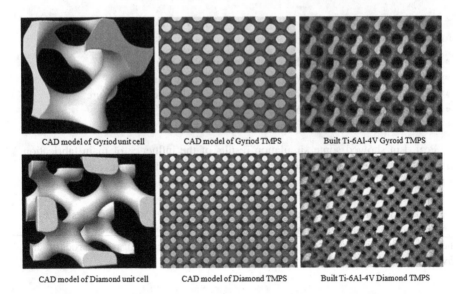

Fig. 3.5 CAD models of the unit cells and TMPS cellular structures, and optical micrographs of SLM-made Ti-6Al-4V TMPS cellular structures

(1) The micro-computer tomography (CT) scan results reveal that the TPMS cellular lattice structures with a wide unit cell size range of 2 to 8 mm and the designed volume fractions as lower as 6% can be manufactured free of defects by the SLM process without the need of support structures. The scanning electron microscopy (SEM) micrographs show that the Gyroid lattice structures made by SLM are well defined and have a good geometric agreement with the original CAD model.

(2) It has been demonstrated that this type of unit cell with the self-supported feature extends the capability of SLM in producing cellular lattice structures with a broad range of unit cell size, which were previous thought difficult or impossible to be made.

(3) The yield strengths and Young's moduli both increase with the increase in the volume fraction of the lattice structures. The equations based on Gibson-Ashby model have been established to use in the future design to estimate the approximate compressive modulus and strength of the SLM-manufactured TPMS cellular lattice structures without mechanical testing.

3.5 Fundamental Research in Additive Manufacturing at BIC-ESAT

[*] Huiling Duan recognized contributions from the following researchers: M. Liu, X. Wei, K. Zhang.

Prof. Huiling Duan provided a review of research progress on advanced additive manufacturing (AAM) technologies for functional and smart materials/structures/systems at multiple scales and under couplings of multi-physic, which is developed by the Beijing Innovation Center for Engineering Science and Advanced Technology (BIC-ESAT) at Peking University. Research at BIC-ESAT focuses mainly on several fundamental scientific interests of Mechanics in Additive Manufacturing, including manufacturing process simulations, microstructure predictions, thermo-mechanical process modeling, etc. By decoupling the complex manufacturing process into steps, the studies on the durability of underwater superhydrophobic surface and its recovery have been carried out (Lv et al. 2014; Xue et al. 2015). The results show an improved behavior of nozzle jetting by modifying the printing head with hydrophobic surface. This finding could be used to improve the efficiency, accuracy of 3D printing process, also the quality of the printed materials. Based on the spreading and infiltration of droplets on rough surfaces, the controlled powder printing and the enhanced mechanical properties of printed materials have also been observed. Besides, there are potential applications in health systems and multifunctional electronic devices by using 3D printing technology, such as metamaterials, microfluidic chip for medical diagnostics and 3D printed micro/nano-robots in bio-medicine, etc. For example, by using nano-fabrication technology, different sizes and shapes of micropumps are fabricated for targeted drug/gene/cell delivery in vivo and vitro. Moreover, a tunable digital elastic metamaterial fabricated by 3D printing technology, which allows programmable broadening of the metamaterial band gap for vibration isolation in many engineering fields have been proposed (Wang et al. 2016).

Prof. Duan introduced that BIC-ESAT has made progresses in simulations of additive manufacturing processes, especially in particle-based methods for modeling powder-based additive manufacturing (Liu and Liu 2016). Among many meshfree, Lagrangian, particle methods, smoothed particle hydrodynamics (SPH) and dissipative particle dynamics (DPD) were chosen because they are easy to deal with large material deformation and convenient in tracking moving material interfaces. Both SPH and DPD have been developed with advanced algorithms and integrated into a code package, which is able to model the melting and solidification of metal, impacting, spreading and deposition of drop onto substrate, and multiphase flow into porous structures. With future developments, the code package is expected to model important physics in AM including powder transport, coupled thermo-mechanical-fluid flow (melt pool dynamics), drop dynamics and many other complex multi-physical processes involved in additive manufacturing.

She also emphasized that BIC-ESAT has made progress in developing continuum theories that connect the microstructures with the mechanical properties of macroscopic composite materials, including: (a) Through continuum mechanical analysis, a statistical shear lag model has been developed to explore the size effects in hierarchical composites' mechanical properties. The model explains why composites with hierarchical structures show very different size dependency in their mechanical properties (Wei et al. 2015), and achieve improved defect-tolerance than homogeneous materials. (b) Experimental methods and techniques have been developed to investigate the critical mechanical properties of constituent materials at multi-level length scales in carbon-based composites. These methods provide the essential data base (e.g. adhesion, strength, etc.) that are important for precise multiscale simulations of high-performance composite materials.

Finally, Prof. Huiling Duan expected that the theoretical, computational and experimental studies at BIC-ESAT will result in a novel design philosophy and innovative manufacturing technologies that ultimately lead to light-weight yet high-performance multifunctional alloys and composite materials.

3.6   Thermo-Mechanical Analysis of Additive Manufacturing Process by Blown Powder Technology

* Co-workers: M. Cervera, W. Huang, X. Lin and L. Wei.

Prof. Michele Chiumenti began by showing the differences between three main technologies of metal deposition for AM:

(1) Wire-feeding technology
(2) Blown powder technology: Laser Solid Forming (LSF)
(3) Powder bed technology: Selective Laser Melting (SLM)

He points out the differences between the *global* level analysis (the structural response) and the *local* level analysis (melting pool). In this work, the numerical simulation of AM process at global level is addressed. A fully coupled thermo-mechanical framework is used to solve the governing equations (balance of energy and balance of momentum) as well as the constitutive laws to describe the material behavior in the entire temperature range.

The proposed simulation strategy is suitable for both wire feeding and blown powder technology. It consists of a high fidelity analysis able to reproduce the exact deposition (scanning) sequence as set for the machine and giving detailed information about the temperature and the stress field during the whole process duration. The simulation strategy can be resumed by the following key-points:

(1) The *Born-dead-elements technique* is used to activate the elements according to the scanning sequence. The elements used to describe the FE domain can be separated into: (i) Inactive elements that are neither computed nor assembled into the solution system; (ii) Activated elements, exposed to the power input along the scanning sequence within the current time-step, either forming a new layer of material or belonging to the melting pool; (iii) Activated elements are

part of the computational domain: those include the supporting structure and the different layers previously deposited.

(2) An *octree-based searching algorithm* is used to look for the elements belonging to the Heat Affected Zone (HAZ): new material layer ready to be activated as well as part of the melting pool. The searching algorithm is also used to update the boundaries of the computational domain, to account for the accurate heat loss through it.

(3) The *Common Layer Interface (CLI) format* is used to define the scanning sequence as for the numerical control of the machine.

(4) An *enthalpy-based formulation* is used to solve the balance of energy equation including the liquid-to-solid phase-change.

(5) The heat losses by convection and radiation through the external surfaces of the domain have been identified as one of the most sensitive mechanism for an accurate simulation of the AM process. These boundary conditions are changing according to the actual definition of the domain according to the sequence of deposition of the AM process.

(6) *Thermo-viscoelastic-viscoplastic* constitutive model is used to characterize the rate-dependent material behavior within the entire temperature range from solid, to mushy or liquid phases. The material data base is *temperature-dependent* according to the temperature range of the manufacturing process (Chiumenti et al. 2016).

Prof. Michele Chiumenti showed a large number of numerical simulations intended to calibrate and validate the proposed model for AM. An extensive experimental campaign has been carried out at the *State Key Laboratory of Solidification Processing (SKLSP), at the Northwestern Polytechnical University of Xi'an* where a LSF machine is used for the calibration of the power input, the heat losses and, particularly, to match the mechanical response of the constitutive model. The LSF machine makes use of the blown-powder technique to deal with the Metal Deposition (MD) process in a layer-by-layer manner. Both the software and the machine read the same scanning sequence given through a CLI format, that is, a sequence of polylines and hatching strategy to fill the entire section. The power absorption coefficient and the Heat Transfer Coefficients (HTC) for both the heat convection and the heat radiation laws have been calibrated to capture the temperature evolution at the different locations where the thermocouples have been placed (Chiumenti et al. 2016). The mechanical response of the thermo-viscoelastic-visco-plastic constitutive model has been calibrated by comparing the distortion of the supporting plate at different locations monitored during the full duration of the manufacturing process for a Wire-feeding process (Chiumenti et al. 2010). Temperature evolution, distortions and residual stresses are shown in Figs. 3.6 and 3.7, where the metal deposition of a hanging-lug for an aeronautical turbine component is presented.

Moreover, Prof. Michele Chiumenti described the simulation strategy adopted for the analysis of the SLM technology (powder-bed method). The main differences compared to the wire-feeding technology are: (i) thinner layer thickness (30–

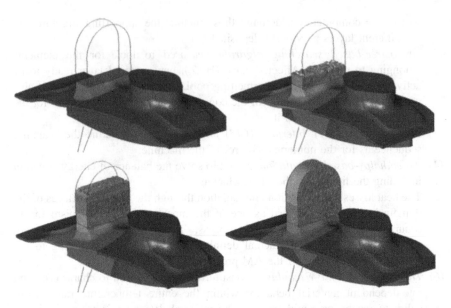

Fig. 3.6 Temperature evolution. *Source* Workshop presentation by Prof. Michele Chiumenti (Universidad Politécnica de Cataluña)

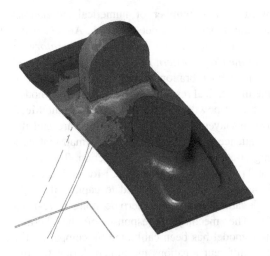

Fig. 3.7 Residual stresses (J2). *Source* Workshop presentation by Prof. Michele Chiumenti (Universidad Politécnica de Cataluña)

60 microns); (ii) faster scanning strategy; (iii) lower power input and correspond-ingly, much smaller HAZ. The simulation strategy proposed to deal with the SLM process is based on the *inherent strain method* (Chiumenti et al. 2017):

(1) The high fidelity scanning strategy is replaced by a *layer-by-layer deposition sequence*;
(2) The *Born-dead-elements technique* together with the Octree-based searching algorithm is used to look for the elements belonging to each new layer.
(3) A pure mechanical analysis is performed to compute the stress distribution induced by the thermal shrinkage of each new layer deposited. A viscoelastic-viscoplastic constitutive model with constant material properties is used to char-acterize the material behavior.
(4) The inherent strains are defined as a function of: (i) the temperature gradient between the melting temperature and temperature of the supporting structure; (ii) by the subjacent scanning strategy adopted.

Hence, this approach accounts for the thermal shrinkage effects during the cooling phase as the main source of distortion of the AM component. Hence, all the elements belonging to the current layer, shrunk from the melting to the room temperature.

Both simulation strategies proposed by Prof. Michele Chiumenti and addressed to the numerical simulation of different AM process have been experimentally calibrated and validated. The remarkable agreement with the experimental evidence achieved demonstrates the good accuracy of the methods proposed.

The concluding remarks are the following:

(1) A remarkable accuracy for the simulation of AM process by metal deposition has been achieved for blown-powder, powder-bed and wire-feeding technologies.
(2) A fully coupled thermo-mechanical framework allows for the high-fidelity simulation of the AM process of both wire-feeding and blown-powder tech-nologies. The actual scanning sequence is reproduced adopting the same CLI input format as for the machine.
(3) In case of powder-bed technology (i.e. SLM) the CPU-time can be reduced using a layer-by-layer activation sequence and adopting the Inherent Strain or Inherent Shrinkage methods.
(4) The thermo-viscoelastic-viscoplastic constitutive model proposed is very suit-able to capture the residual stresses and the distortions induced by the heat source during the sintering process.
(5) An exhaustive experimental campaign to calibrate and validate the proposed numerical models has been carried out with the support of different EU projects.

The on-going and future developments of the research group at CIMNE are the following:

(1) Development of a massive parallel HPC software platform operating in both shared and distributed memory machines, allowing for large scale industrial computations.

(2) Continuous self-balancing domain decomposition strategy suitable for the building operation typical of AM processes.
(3) Octree based adaptive meshing allowing for local refinement (melting pool) and global coarsening.
(4) Embedded formulation allowing for the use of complex CAD geometries immersed into background voxel meshes.

3.7   Challenges in Macroscopic Modeling of the PBF Process

Prof. Lars-Erik Lindgren began by showing a figure of IDeoM^{2}P^2 (see Fig. 3.8), and explained that AM is a meeting place for the concepts: Microstructure evolution, Computational thermo-mechanics, manufacturing processes, Performance and lifetime, Component performance.

Then he reviewed the research of Computational Welding Mechanics (CWM) and Weld Process Models (WPM). He stated that the past of CWM gives several indications for the future of Predictive Theoretical, Computational and Experimental Approaches for AM.

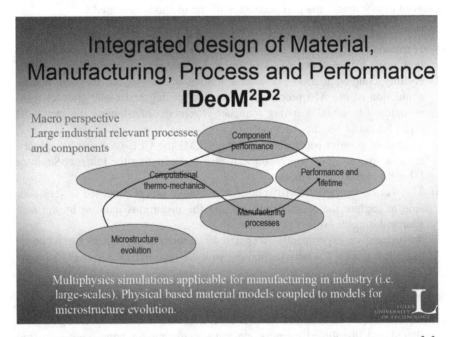

Fig. 3.8 Integrated design of material, manufacturing, process and performance (IDeoM^{2}P^2). *Source* Prof. Lars-Erik Lindgren, Luleå University of Technology, presentation to the workshop

However, there are also some different conditions making the current state and future outlook better than when CWM started to develop in the 1970s, according to Prof. Lindgren. The difference is not only the enormous development in CWM and other modeling areas together with computer hardware and software since that time. It is also today's large interest from science, engineering and business in AM. He noted that welding was considered an old technique when CWM developed and the interest from scientific community and industry for developing predictive models was relatively limited.

There is a difference between CWM and WPM, he stated. The aim in CWM is to predict the overall behavior of a welded component, during welding as well as final properties. Its focus is on the domain outside the weld region and replaces that physics with a heat input model and focuses on the larger scales. It is typically solved by Finite Element Methods. WPM is a modeling approach where the focus is on the weld pool region. The required modeling approach often includes multi-physics and a variety of numerical methods applicable for that scale and field equations used. CWM models, on the other hand, starts with a given heat input that replaces the details of the heat generation process and focuses on the larger scales. Therefore, the CWM and WPM approaches are complementary. These concepts carry over to simulation of AM processes.

Prof. Lindgren then showed some recent research of AM, for example, the talk described the application of CWM approaches for macroscopic modeling of AM processes (Lindgren et al. 2016). Simulations of directed energy deposition (DED) processes were demonstrated for Ti-6Al-4V (Lundbäck and Lindgren 2011), see Fig. 3.9, and for Alloy 718 (Fisk et al. 2012). A dislocation density based flow stress model was coupled with a phase change model in the first case. It was coupled with a precipitate growth/dissolution model in the Alloy 718 case.

Furthermore, the talk discussed modeling of powder bed fusion (PBF) processes where the variation in length scales and the amount of 'welds' is very large. In AM each layer has a height around 0.1 mm. This means that building a 2 dm long wall with 5 cm height has 100 m of 'welds'. This requires various modeling

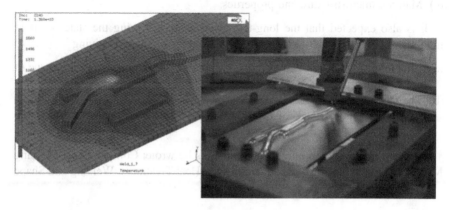

Fig. 3.9 Validation case for AM of Ti-6Al-4V

simplifications combined with computational strategies that can enable a sufficient accurate model. Thus it is numerically more challenging to simulate PBF than DED processes, Prof. Lindgren commented.

The simplification methods for reducing the computational efforts, same as used in the early decades of CWM, are summarized with a particularly focus on the lumping approach. Evaluation of the latter approach shows that a speed up of more than 25 times can be achieved with reasonable accurate temperatures and deformations. The author summarized the prospects for the future at the end of his talk. AM is a meeting place for development of modeling, numerical methods, materials and process in a similar way that has happened in CWM with focus on welding.

Prof. Lindgren noted that this will not only contribute to Production Technology (development of the AM process and products) but also to Materials Science (developments of materials) and Computational Mechanics (development of multi-physics models and software). CWM in welding leads to predictive models for deformations, stresses, microstructure and some indicators for crack initiation. However, practices for life prediction never got established and many critical applications required full scale testing for qualification.

He stressed that the future seems brighter for predictive models for the AM arena in this respect, as much more resources and a large variety of expertize are involved in the future development of CWM and WPM approaches for AM processes. Then Lindgren raised some issues to be developed. The expected short-term (five years) issues are within education;

(a) Transforming engineering design by merging materials, process and product design.
(b) Eliminating its dichotomy into materials science, production and mechanics as well as restructuring engineering education.

A larger design space enabling innovative engineering solutions will be possible by integrated design of

(a) Product performance,
(b) Manufacturing process physics,
(c) Material microstructure and properties.

It is also expected that the long-term (10 years) will bring the state of art in modeling of AM beyond the state of art in welding simulations giving

(a) Coupling of WPM and CWM models where the former prescribes the shape of the solidified material to CWM models.
(b) The prediction of material state (deformation, microstructure) also includes defect state (pores, lack of fusion) enabling process improvement and more rapid qualification procedures.

At last, Prof. Lindgren introduced the books he wrote: Computational Welding Mechanics, Encyclopedia of Thermal Stresses: Heat Treatment, Welding and Shape

Memory Materials, Finite element simulation of manufacturing, Plasticity and microstructure models for metals and alloys (in preparation).

3.8 Computational Modelling of Heat Generations and Microstructural Evolutions in Advanced Manufacturing Technology

[*] Zhao Zhang recognized contributions from the following researchers: Peng Ge (Dalian University of Technology), Zhenyu Wan (Dalian University of Technology), Chaoping Hu (Dalian University of Technology), Zhijun Tan (Dalian University of Technology), Qi Wu (Dalian University of Technology), Jingping Liu (Dalian University of Technology).

Prof. Zhao Zhang gave a short introduction about the collaboration from his research group with international scholars on Monte Carlo simulations and the phase field modelling of micro-crack propagations. Then, he introduced the mainly developments of heat source calculations in the field of laser additive manufacturing (LAM). Based on the mentioned literatures (Manvatkar et al. 2014), they proposed a new heat source model for LAM in which the particle sizes and numbers can be explicitly expressed. Some examples are given to show the effect of particle sizes and numbers on the temperature histories. Melting and re-melting phenomenon can be found from the calculated results (Fig. 3.10) and Prof. Zhang pointed out that the melting and re-melting phenomenon can affect the following simulations on grain growth in additive manufacturing.

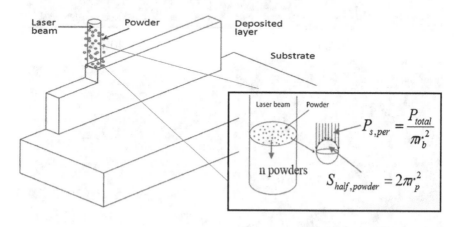

$$q(r) = \frac{6\sqrt{3}\eta P_{absorb,bed} f}{abc\pi\sqrt{\pi}} \exp\left[-\left(\frac{3x^2}{a^2} + \frac{3y^2}{b^2} + \frac{3(z+vt)^2}{c^2}\right)\right] = \frac{6\sqrt{3}\eta P_{total} f}{abc\pi\sqrt{\pi}} \exp\left[-\left(\frac{3x^2}{a^2} + \frac{3y^2}{b^2} + \frac{3(z+vt)^2}{c^2}\right)\right] \cdot \left(1 - \eta_z \cdot \frac{n \cdot 2r_p^2}{r_b^2}\right)$$

Fig. 3.10 Heat source model

From the sequentially coupled thermo-mechanical model for additive manufacturing, they obtained the thermal stress histories and the residual stress. They used the Crossland criterion to determinate whether a micro-crack can be formed from a void defect. Then, they transformed the void defect into an equivalent micro-crack. A simple example is given to show the effect of the thermal stress history on the possible micro-crack propagation by phase field method.

He then introduced the Monte Carlo model for grain growth in additive manufacturing for Ti-6Al-4V alloy. The two phase grain growths for Ti-6Al-4V in laser welding/additive manufacturing and in friction stir welding/friction stir additive manufacturing are simulated. In comparison with the changes of volume fractions of α/β phases, they validated the proposed model. They found that with the decrease of the temperature in the cooling stage, α is formed on the β substrate. The volume fraction of α is increased in the cooling. In air cooling or higher cooling conditions, the lamellae or needle grains can be formed. The lamellae α grain generates on the boundaries of the β grain, and then grow into the interior of the beta grains (Fig. 3.11).

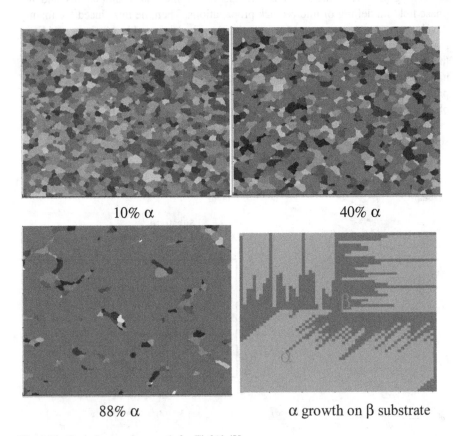

10% α   40% α

88% α   α growth on β substrate

Fig. 3.11 Two phase grain growth for Ti-6Al-4V

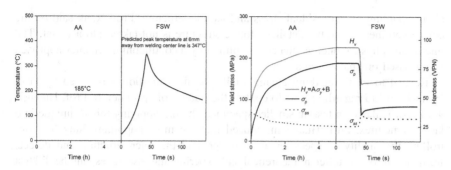

Fig. 3.12 Temperature history and predicted mechanical property

Prof. Zhang introduced the precipitation evolutions in friction stir welding and then showed the effect of the precipitation on the grain growth in friction stir welding process. They introduced that the precipitation kinetics of Al–Mg-Si super saturated solid solution during artificial ageing process: SSSS → Atomic clusters → G.P.-zones → $\beta''(Mg_5Si_3)$ → $\beta'(Mg_5Si_3)$ → $\beta(Mg_2Si)$. Based on the rate law and the solute interface concentration reduced by the Gibbs-Thomson equation, the critical radius r^* during evolution can be obtained to determine whether particles of different sizes grow or dissolve in the matrix according to the Ostwald ripening. They divided the yield stress into three parts: the contribution from pure metal, the contribution from the solid solution and the contribution from the second phase. KWN model was used for the calculation of precipitations. They can then predict the yield stress of Al–Mg-Si alloy from different temperature conditions (Fig. 3.12).

He proposed several possible tasks related to current modelling: (1) Although Monte Carlo method can be used to simulate the grain growth in single phase and two phase materials, computational methods have to be further developed to simulate different microstructural evolutions in different material systems in LAM. (2) Thermal stress cycles and residual stresses of LAM have been simulated. The effects of the obtained data on the fatigue properties with consideration of defects need to be further studied. The determination of the threshold values for the imperfections can be useful for the online monitoring and offline monitoring in LAM. (3) Although the mechanical properties of Al–Mg-Si alloy can be accurately predicted, more efficient method has to be developed to predict the mechanical properties of different material systems.

3.9 Discussion

After Prof. Rank's talk, the first question was posted regarding the computational cost about the finite cell method. Prof. Rank noted that they did a lot of numerical experiment. If implementing it at a right way, it is very fast. Another question was about coupling of the residual stress and atomic structure. Prof. Rank said on the considered meso-scale the detailed atomic structure is not taken into account. The microscale physical phenomena are simplified by a continuum mechanics approach being based on spatial averaging.

Related to Prof. Weidong Huang's talk, the first question was posed by Prof. Wing-Kam Liu regarding how to control the quality of specimen in Prof. Huang's work, particularly on testing smaller specimens before applying this technique into larger structures. Prof. Huang introduced that the most important thing for controlling the quality of specimen is to understand the microstructure and defects formation through sufficient theoretical and experimental researches, and he did test a lot of specimens' quality in smaller domain and also in large size part. The second question was posed by a participant on how to deal with different kinds of materials. Prof. Huang explained that firstly some same basic methods could be applied for different materials in theory, while parameters must be different for different materials in practice, for example Ni or Ti alloy. Another participant asked Prof. Huang on his comments on the surface finish. He thought that the surface finish was not that necessary in his work and other post processing methods were used. Finally, a participant asked Prof. Huang regarding how to consider the mesh anisotropy in the CA modeling of additive manufacturing. Prof. Huang explained that eliminating the mesh anisotropy is very important in CA modeling of additive manufacturing, and recommended that some special work on this problem has been made by Dr. Wei in his group.

Related to Prof. Feng Lin's presentation, the first question posed by Levine was that, focusing on the very large changes of composition due to the evaporation, where are all the evaporated materials and how they go back on to the surface in the numerical simulations. Prof. Lin explained that the computational elements are distributed at everywhere and the surfaces are travelled during the numerical simulations. The second question was posed by another participant. The participant required some comments on that the particles come off the bed and this becomes more and more diffuse at the state. He commented that the EBSM is different from the laser and the particles would bounce out everywhere when the particles hit the powder bed. So, it is very hard to explain how this happen. On the other hand, the EB scans the powder bed and the particles are limited to run out of the powder bed.

Concerning Prof. Chunze Yan's talk, the first question was posed by a participant regarding suggesting Prof. Yan to do statistics analysis for data obtained in simulation, in order to make the numerical model more precisely. Prof. Yan agreed that. The second question was posed by a participant regarding how small of structure in Prof. Yan's work, as the participant think it is quite hard to obtain this kind of structure when the scale is small. The third question was also concerning in the size of structures. Prof. Yan explained that his work was not in the nanoscale at the moment. The last question was posed by a participant concerning that the test made in Prof. Yan's work could be extended into more kinds of applied conditions, instead of only bending dominant conditions. Prof. Yan agreed this too.

Related to Prof. Huiling Duan's work, the first question was posed by chairman of this session Prof. Jun Yan regarding the advantage the honeycomb structures.

Prof. Duan explained that she has tried to use natural honeycomb structures and find this kind of structures has advantages as they had natural fiber-reinforced structures inside. The second question was posed by a participant regarding how to supply power for micro-robots. She explained that this can be made by using a kind of traditional machine. The last question was posed by another participant regarding how precise by using inkjet technique in her work. Prof. Duan introduced that she has tested this in her work and she provided a data obtained in this experiment.

Referring to Prof. Michele Chiumenti's talk, a first question regarded the FE technology used to deal with the incompressibility behavior of the liquid and mushy phases as well as to deal with the purely deviatoric nature of the visco-plastic strains. The answer has clarified that the software makes use of both mixed displacement/pressure formulation or Q1P0 elements, suitable for those conditions. A second question referred to the meshing strategy. The answer has shown the use of a coarse mesh far away from the HAZ to concentrate it at the melting pool where the sintering process takes place. Hence, a mesh optimization is required to optimize the CPU-time.

Related to Prof. Lars-Erik Lindgren's work, a question was regarding the fluid flow and phase change in the molten zone. Prof. Lindgren noted macro modeling replaces the details of the heat generation process or phase change in this region and focuses on the larger scales.

Regarding to Prof. Zhao Zhang's talk, the first question from Prof. Lindgren is focused on the selections of the parameters in the prediction of mechanical properties of Al–Mg–Si alloy. They should be temperature dependent. Prof. Zhang responded that the purpose of the current selection is for validations. So, they have to be selected to be exactly the same in the used literatures for comparisons. In the next step, the parameters should be selected to be temperature dependent for more accurate description on this point. The second question from Prof. Rollett is mainly about α phase growth. Prof. Zhang responded to this question and stated that the α phase is growing mainly along the given 110 direction when the cooling rate is high. For smaller cooling rate, it is more likely to obtain equiaxed grains for Ti-6Al-4V. The last question is coming from Prof. Bi Zhang. He asked how to validate the obtained residual stress. Prof. Zhang responded that the residual stress for welding has been widely validated. But for the additive manufacturing, more work need to be done due to the lack of sufficient experimental data.

Chapter 4
Theory, Methods and Tools for Additive Manufacturing Oriented Design and Optimization

The fifth and sixth sessions of the workshop provided an overview of the design in additive manufacturing (AM), including topology optimization and various design approaches in industry. Prof. Ole Sigmund (Technical University of Denmark, Denmark), Prof. Xu Guo and Prof. Shutian Liu (Dalian University of Technology, China), Prof. Michael Yu Wang (Hong Kong University of Science and Technology, China), Prof. Pinlian Han (Southern University of Science and Technology, China), and Prof. Xiangming Wang (Shenyang Aircraft Design and Research Institute of AVIC, China) each discussed researches, challenges, and future directions related to this topic.

4.1 Topology Optimization and Additive Manufacturing—An Ideal Marriage

A general introduction to topology optimization and recent applications with regards to AM methods was provided. Special emphasis was given to optimal infill design that provides increased buckling resistance and robustness towards varying load cases. The relationship between topology optimization and AM was compared to an ideal marriage, means that the two methods are complementary to each other. It was pointed out that topology optimization is a free material distribution approach that provides optimal geometries for specific objective functions and constraints (Groen and Sigmund, 2016; Sigmund, 2011; Sigmund and Maute, 2013; Sigmund, Aage, and Andreassen, 2016; Wang, Lazarov, and Sigmund, 2011; Aage, Andreassen, Lazarov, and Sigmund, 2016). AM methods enable rapid manufacturing of complex geometries. Then they do constitute an ideal marriage. Nevertheless, there are still a number of challenges need to be overcome in order to combine them, such as overhang constrains, support cost, infill, residual stresses, anisotropies, etc.

© The Author(s) 2018
X. Guo et al., *Report of the Workshop Predictive Theoretical, Computational and Experimental Approaches for Additive Manufacturing (WAM 2016)*, SpringerBriefs in Applied Sciences and Technology, https://doi.org/10.1007/978-3-319-63670-2_4

The topology optimization method was first proposed by Bendsøe and Kikuchi. Since its introduction, the topology optimization method has undergone a tremendous development and has been applied to many engineering and science areas, shown in Fig. 4.1. The developments have spread in many different directions as reviewed in (Sigmund and Maute, 2013), such as density approach, topological derivatives and level set approach, etc. Now the method has become an everyday design tool in major industries throughout the world. There exist a number of specialized commercial codes for topology optimization and most finite element software packages also provide simplified approaches.

The "TopOpt App", which is developed by the TopOpt group at DTU (www.topopt.dtu.dk), was then introduced to the workshop. These real-time interactive Apps can run on tablets and mobile phones (download in AppStore (IOS) and Google Play (Android), the ultimate goal of these Apps is to give users real-time Giga-resolution. A first in achieving Giga-voxel design resolution for full-scale airplane wing design has also been demonstrated by the TopOpt group (Aage, Andreassen, Lazarov and Sigmund, 2016).

It is pointed out that, a popular problem at present is how to make the adaption of topology optimization to additive manufacturing methods. However, a lot of challenges need to be overcome such as overhang constrains, support costs in topology optimization and overhang limitations. Citing the words of Sigmund is "In a perfect marriage partners adapt to each other flaws!" The TopOpt group has

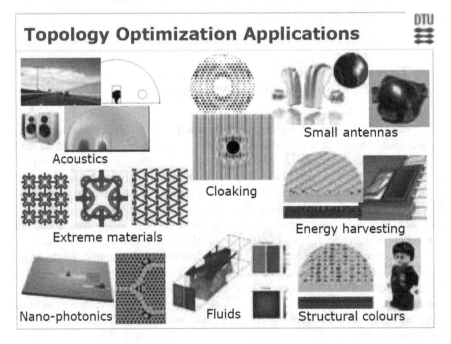

Fig. 4.1 Topology optimization applications. *Source* Prof. Ole Sigmund, Technical University of Denmark, presentation to the workshop

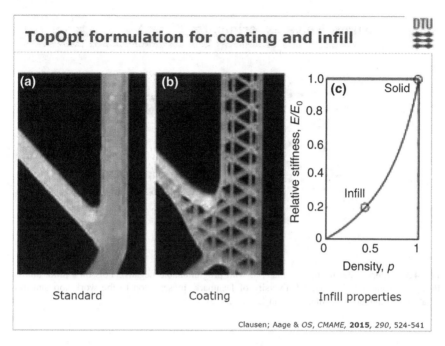

Fig. 4.2 TopOpt formulation for coating and infill. *Source* Prof. Ole Sigmund, Technical University of Denmark, presentation to the workshop and also (Clausen, Aage and Sigmund, 2016)

concentrated on infill design (Clausen, Aage and Sigmund, 2015; Clausen, Aage and Sigmund, 2016; Wu, Aage, Westermann and Sigmund, 2016), taking advantage of the Additive Manufacturing methods' ability to generate structures with internal microstructure and in turn achieving increased structural stability (Clausen, Aage and Sigmund, 2016). Constant density porous isotropic infill does not increase stiffness compared to solid designs (Clausen, Aage and Sigmund, 2015; Clausen, Aage and Sigmund, 2016) as shown in Fig. 4.2. And likewise, open-walled lattice structures are sub-optimal with respect to stiffness (Sigmund, Aage and Andreassen, 2016).

Then some examples for application of topology optimization using AM method were introduced. One is the design of extreme material with programmable negative Poisson's ratio with manufacturing constraints allowing them to be efficiently built using AM soft polymer extrusion processes (Clausen, Aage and Sigmund, 2015). Another is applying topology optimization to the design of metallic passive cooling devices driven by natural convention. Finally, some experimental testing results were presented which have verified the theoretical results and prove that cooling structures with unprecedented performance can be achieved with the combined topology optimization and AM process (Alexandersen, Sigmund and Aage, 2016; Alexandersen, Aage, Andreasen and Sigmund, 2014), shown in Fig. 4.3 and Fig. 4.4.

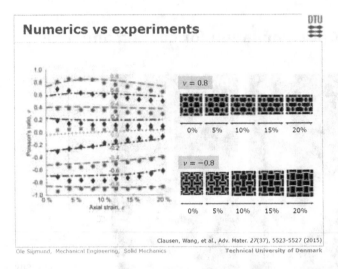

Fig. 4.3 Experimental results for material with programmable negative Poisson's ratio. *Source* Prof. Ole Sigmund, Technical University of Denmark, presentation to the workshop and also (Clausen, Wang, Jensen, Sigmund and Lewis, 2015)

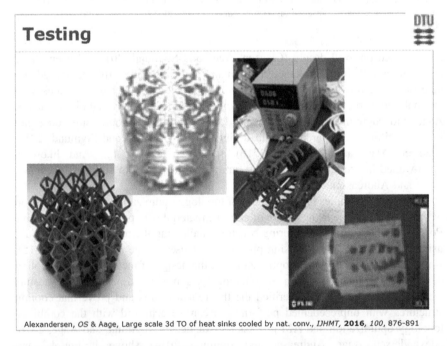

Fig. 4.4 Experimental testing for metallic passive cooling devices driven by natural convention. *Source* Prof. Ole Sigmund, Technical University of Denmark, presentation to the workshop

In conclusion, there is no doubt that topology optimization and additive manufacturing constitutes an ideal marriage. However, as is the case in all relationships, partners must adapt to make things work. So far, a lot of work has been done on the topology optimization side in order to adapt to AM limitations. However, with the improvement of AM methods, some parts of this work may turn out to be wasted and new interesting challenges may arise as the couple develops.

4.2 When Additive Manufacturing Meets Topology Optimization: Current Status and Challenges

AM is a relatively new manufacture technique which enables the fabrication of components in an additive (layer-by-layer) way. On the other hand, topology optimization, which aims at designing innovative and lightweight products by distributing material within a prescribed domain in an optimal way, has reached a certain level of maturity and becomes a well-established research area. Although topology optimization has great potential to become a perfect design tool that can fully exploit the tremendous design freedom provided by AM, it must be admitted that existing topology optimization approaches cannot be fully adapted to the current AM techniques. Prof. Guo gave a summarized brief of limitations which must be taken into account.

Then Prof. Guo introduced that his presentation would focus on how to design structures that are self-supporting in a manufacture process without introducing additional supporting materials, which is one of the challenging issues in AM oriented topology optimization. Compared to adding extra material or carrying out post-processing to make an optimal structure printable, it is generally believed that designing self-supporting printable structures through topology optimization method directly is more preferable since it can simplify the post-processing and reduce the manufacture cost. His motivation is from the consideration that since the constraints associated with the designing of self-supporting structures (e.g., overhang angle, minimum length scale) are actually geometrical in nature, it seems more appropriate to include more geometry information in the problem formulation and perform topology optimization in a geometrically explicit way.

He explained there are two solution approaches established based on the Moving Morphable Components (MMC) and Moving Morphable Voids (MMV) frameworks that can be used to solve the problem such as this: (1) under the MMC-based explicit topology optimization framework, the only extra constraints is $(\sin(\theta_k + \alpha))^2 \geq (\sin(\bar{\theta}))^2, k = 1, \ldots, nc$ where θ_k is the inclined angle of the k-th component, α is the rotation angle of the work plane and $\bar{\theta}$ is the lower bound of the overhang angle, respectively (see Fig. 4.5 for reference). (2) Under the MMV-based topology optimization framework, a set of printable voids with explicit boundary representation as the basic building blocks of topology optimization as shown in Fig. 4.6. In

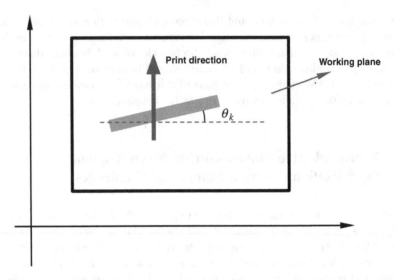

Fig. 4.5 The construction of a B-spline curve based printable feature. *Source* Profs. Xu Guo and Shutian Liu, Dalian University of Technology, presentation of Prof. Guo to the workshop

order to preserve the printability of each void in the structure, it is required that every two voids cannot be intersected otherwise the printability of the resulting void may not be guaranteed.

In his presentation, some theoretical issues associated with AM oriented topology optimization were also stated. Prof. Guo analyzed that the theoretically optimal self-supporting structure can be constructed from the corresponding optimal structure without considering the self-supporting requirement by introducing infinitely many rods with infinitely small cross sections as additional supporting structures. If no regularization technique is introduced (e.g., imposing minimum length scale constraint or total perimeter constraint, etc.), it can be expected that the numerical solution results may be highly mesh-dependent if the solution algorithms are smart and robust enough to find true global optimal solutions. Of course, if regularization formulations/techniques are introduced in prior in problem formulation or employed in the numerical solution process, these "chattering designs" can definitely be suppressed. Although the above analysis is only theoretical in nature, it provides useful insight into the problem under consideration and gives estimation on the lower bound of the optimal value of the objective functional.

Furthermore, in order to control the minimum length scale in topology optimization, a problem formulation for length scale control in optimal topology designs was also proposed by Prof. Guo. He explained that the concept of medial surface is adopted in his approach for length scale control. Compared with the existing length scale control approaches, the advantages of his approach are such that it is a local and explicit length scale control scheme and can generate pure

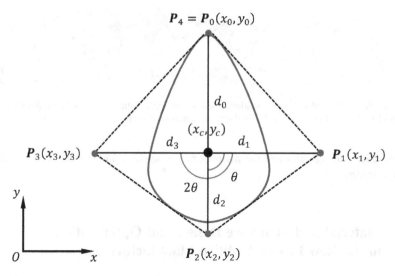

Fig. 4.6 The construction of a B-spline curve based printable feature. *Source* Profs. Xu Guo and Shutian Liu, Dalian University of Technology, presentation of Prof. Guo to the workshop

Fig. 4.7 The optimized structure without considering self-supporting requirement. *Source* Profs. Xu Guo and Shutian Liu, Dalian University of Technology, presentation of Prof. Guo to the workshop

black-and-white designs without resorting to any post-processing or continuation treatment.

Then, Prof. Guo provided several numerical examples to demonstrate the effectiveness of his approaches for designing self-supporting structures through topology optimization. As shown in Fig. 4.7 and Fig. 4.8, the optimized designs obtained by his approaches is purely to satisfy the printability requirement.

Prof. Guo emphasized that: (1) topology optimization framework for AM still needs further explorations. (2) Multi-scale optimization is a bottleneck for when AM oriented topology optimization approaches. (3) Multidisplinary optimization considering the AM process is still a challenging problem.

At last, Prof. Guo introduced The International Research Center for Computational Mechanics at Dalian University of Technology, which is established based on the Department of Engineering Mechanics and the State Key Laboratory of Structural Analysis for Industrial Equipment of Dalian University of Technology. The center is oriented to the international academic frontier in

Fig. 4.8 The optimized structure considering self-supporting constraint. *Source* Profs. Xu Guo and Shutian Liu, Dalian University of Technology, presentation of Prof. Guo to the workshop

computational mechanics and the national major demand for scientific computation in engineering.

4.3   Material and Structure Design and Optimization in the New Era of Additive Manufacturing

Prof. Michael Yu Wang presented some new works done in his group, which is related to an effective cellular structural design methodology for AM technology. Firstly, he pointed out that emerging opportunities are coming for the combination of topology optimization and additive manufacturing. World-wide markets are tremendous for this field, due to the fact that the structures with better mechanical

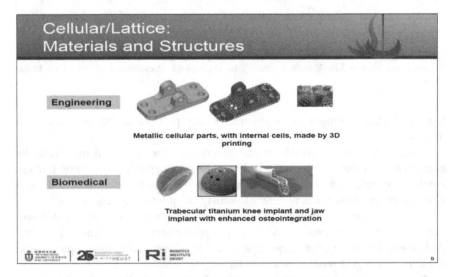

Fig. 4.9 Applications for cellular structures. *Source* Prof. Michael Yu Wang, Hong Kong University of Science and Technology, presentation of Prof. Wang to the workshop

performance are always geometrically complex and they can be created via the flexible AM technology, as shown in Fig. 4.9.

Secondly, he stated that design of both the global structural geometry and the downscaling micro-architectures is vital to create superior-performance and multifunctional structures. In particular, rapid advance of AM releases a great potential to fabricate innovative shape-complexed structures, leading to increasing attention to the multiscale design field. However, Prof. Wang said that there still exist two major issues typically encountered when performing a multiscale design process. One is the computational cost related to designing a huge number of microstructural candidates. The other more critical issue is the lack of the connectivity between the spatially-varying microstructures, which is mainly caused by the decouple design strategy. These unconnected microstructures may lead to impending member parts, which cannot carry any external loads and are also impossible to be produced. Although some improved methods have been claimed to tackle the issue, the complexity of the optimization problem is not inherently reduced.

Thirdly, he introduced a number of works been done recently in their group towards developing an effective cellular structural design methodology. Both the spatially-varying microstructure configurations as well as their distribution patterns are optimized. Some published works include,

(1) They have developed an integrated design framework in the framework of level sets. In this design framework, both the configurations of the global structure and a periodically-distributed unit cell are optimized. He gave us optimized solutions for a cantilever beam problem which is presented in Fig. 4.10. An important conclusion can be made by comparing these results, that is whether to allocate more base materials to the material cell or to assign more fully solid materials to the

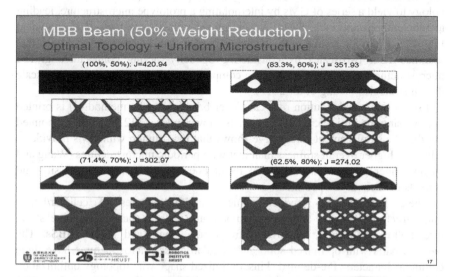

Fig. 4.10 Optimized solutions of MBB beam by method (1). *Source* Prof. Michael Yu Wang, Hong Kong University of Science and Technology, presentation of Prof. Wang to the workshop

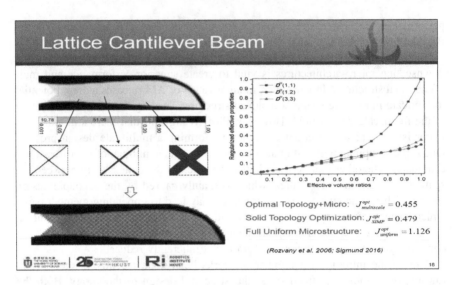

Fig. 4.11 An optimized refine-scale structure by method (2). *Source* Prof. Michael Yu Wang, Hong Kong University of Science and Technology, presentation of Prof. Wang to the workshop

global structure is problem dependent (Wang and Wang, 2016). More complex microstructures can give better mechanical performance.

(2) An efficient concurrent multiscale design method has been developed by considering spatially-varying connectable graded microstructures (GMs). The optimization that spans the two-scale levels seeks both the best GMs configurations and the best GMs distribution. And they developed a shape metamorphosis technology to yield a series of GMs by interpolating a prototype microstructure, leading to inherently connectable GMs. An optimized refined-scaled result is shown in Fig. 4.11. From the results in the picture it is seen that the proposed approach that optimize both two-scales by considering spatially-varying connectable graded microstructures improved performance compared with those results from one-scale design strategies (Wang et al. 2017).

(3) An optimized solution of a cantilever beam using the method (2) is printed out by using the Projet 5500 3D printer, as shown in Fig. 4.12. Prof. Wang pointed out that the displacement-force curve shows that the solution with GMs provides a relatively higher overall stiffness than that with periodic microstructures (Wang and Wang, under preparation). And that result does demonstrate the necessity and benefits of their developed method.

Then, Prof. Wang provided some beautiful pictures of cellular/lattice microstructure made by additive manufacturing method such as Kagome structure, hollow octet-truss, negative stiffness honeycomb and graded BBSs. The metallic parts with optimized solid and cellular/lattice microstructures have three common characters: Pre-defined skins, optimized shape with topology optimization

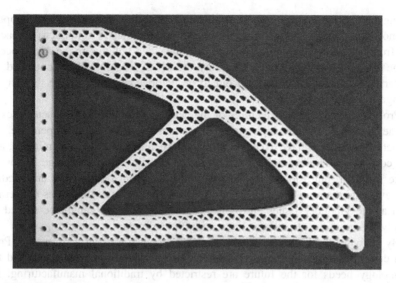

Fig. 4.12 A 3D printed optimized structure with internally periodic microstructures. *Source* Prof. Michael Yu Wang, Hong Kong University of Science and Technology, presentation of Prof. Wang to the workshop

and optimized internal cellular/lattice structures. They have the ability of lightweight, high stiffness, high strength or tailored functionality.

Finally, he concluded that an efficient design methodology has been developed for the multiscale optimization problems. Both the simulation results and the experiment solutions demonstrate that the approach can greatly improve the overall stiffness of the macrostructure. In the future, the proposed method is expected to handle more complicated concurrent design problems, such as by considering the stress-constrained and nonlinearity performance.

4.4   Additive Design for Additive Manufacturing – Challenge from Aero Engine

Prof. Pinlian Han began by introducing himself because he defines himself as an aero engine engineer which he believes is very different. Most of Prof. Han's time has been spent in aero engine business. He went around the world, from China to Germany, Canada, U.S., and back to China. His objective is to make an engine better.

Prof. Han then explained why people love AM. People trust AM because it is traceable and repeatable compared with the traditional manufacturing, according to Prof. Han. Taking forging as an example, it is impossible to repeat a defect caused during the forging process, which can affect the reliability of the parts. This happens for most traditional manufacturing. However, for AM, it starts from micro to macro, i.e., from partial to entire, from point to line, area, and volume. In the AM process, if one can control the very beginning, then the whole process will be controlled, no matter how the end products turn out to be. People know they have the way to change it and to make it better.

Prof. Han showed some pictures of the aero engines. Special characteristics of the aero engines include high speed, high pressure, and high temperature. He expects that they can have excellent properties such as low risk, low emission, and low cost. For a Ti-alloy hollow fan blade, he explained, the challenge is how to make it lighter, but a lot of special requirements except light weight are needed. However, there are no existing materials which can solve all the issues concerned such as the vibratory issues and noise issues. These issues cannot be solved by simply changing the structure or increasing the stiffness.

By showing the life cycle from design to application of an aero engine, Prof. Han commented that the inter-related requirements make it complicated, and the technology needs for the future are restricted by traditional manufacturing. He believes that the solution is additive design and manufacturing. Additive design requires that AM is utilized for bringing unique features. Gradient material and hierarchy structure are the most special characteristics for AM relative to traditional manufacture. Application of them will be the disruptive technology which will change our mechanical world. The best target to apply these new ideas is aero engine, as shown in Fig. 4.13, because of the extreme working conditions as well as

The Solution:
Additive Design and Manufacturing

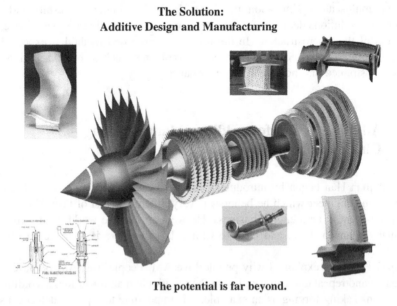

The potential is far beyond.

Fig. 4.13 Aero engine parts. *Source* Prof. Pinlian Han, Southern University of Science and Technology, presentation of Prof. Han to the workshop

the critical safety requirements for the engines. The design for manufacturing for function and property will be changed to design for customer, beyond function and property. Optimization by concurrent AM will be done instead of approaching by iteration. The future design will be based on what people need, not what people can make. Prof. Han also listed the technology needs for tomorrow, including those for compression, combustion, controls, turbines, materials, modeling, mechanical components, and Inlet/Nacelle/Nozzle.

Prof. Han noted that AM is a big deal right now at the General Electric Company (GE). There are lots of news since mid-2011. He quoted some statements from GE, for example, "Within our lifetime, at least 50% of the engine will be made by AM", "In four decades, we could be printing the entire engine", and "Complexity is free". For a LEAP fuel nozzle, only one part is manufactured by AM at GE, compared with the traditional manufacturing, which make it 25% lighter and 5x more durable.

Most people are focusing on material and process such as scan strategy, quantifying variability in AM components, and microstructure-aware modeling from materials processing to mechanical performance. However, Prof. Han stated that those are just evolution, not revolution, and that people's thinking is still in traditional way. When talking about the differences between science and engineering, he raised some issues one should consider, including micro vs. meso vs. macro, material property vs. structure performance, manufacturing processed vs. structure design, and topology vs. sub-surface porous structure. He believes that the meso scale structure is important and simulation of additive design and manufacturing is interesting. He said that the game change includes integration of manufacturing process and field application and combining the modeling process with AM process, and that the behavior in the field application can be tracked back to the manufacturing. The innovation will include new finite element, supper element, implementing the parameters of manufacturing process, such as temperature, size and shape of melting pool, into intuitive equation, big data and knowledge enabling, as well as new methodology and standard for testing and validation, he said.

As for simulation of additive design and manufacturing, Prof. Han proposed some new ways of modeling. One is not dividing an existing structure into finite element but building up the structure by the same path as it is built from AM. Another one is not transferring an existing physical mode to math model but transferring both the physical mode and physical process together into a math model. He emphasized that model building process should be the same as the AM process.

Prof. Han commented that what people need for a structure are reliable design, reliable manufacturing, reliable analysis, reliable test, reliable validation, reliable safety, reliable dependability, and reliable cost. However, so far there have been no tools for design and analysis, no methods for test and validation, no standards for guideline and procedure, and no professionals in this prominent uncultivated land. To be one of the expeditions at Southern University of Science and Technology to explore the dream land, Prof. Han said, he and his collaborators are establishing a new discipline named as "Sub-surface Fine Web Structure Engineering

New Discipline of Engineering
A Challenge to Material, Mechanics, Math and Physics

Sub-Surface Porous Structure
Engineering Mechanics

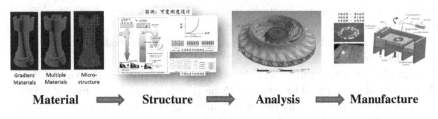

Material ➡ **Structure** ➡ **Analysis** ➡ **Manufacture**

Theory, Methodology, Tool, Standard, Procedure

➡ **Application**

Fig. 4.14 Need for a new discipline. *Source* Prof. Pinlian Han, Southern University of Science and Technology, presentation of Prof. Han to the workshop

Mechanics", as shown in Fig. 4.14. It is going to address all four topics of this workshop. Driven by the concrete requirements for a better aero engine, Prof. Han has tried to throw out a brick to attract a jade, as he added.

To summarize, inspired by nature, the light weight structure with adaptive properties of objective oriented distribution of density, stiffness, vibratory mode, failure mode, damping, heat expansion, etc., are available by proactive additive design and manufacturing. In addition to studying the material property and the physics behind the process of AM, one shall also pay attention to the way of designing a brand-new structure that can only become available due to AM. In addition to studying and understanding the relationship of material property, such as strength and fatigue life, and the process related parameters, such as power, temperature and path, one shall also study how to reduce the stress of a structure as well as the alternating load, including both amplitude and special mode, that cause the fatigue, Prof. Han added. In other words, people shall not only try to enhance the capability of the material to stand high stress but also shall try to arrange the load path to reduce the load gradient that causes high stress by taking advantage of AM. People should not stay at the physics area only to play with those micro behaviors related with AM; people shall encourage the study of the bulk property that can be controlled to display a stable and consistent result.

Additive manufacturing has two key words, according to Prof. Han. The word "manufacturing" is the action to make a part, and the word "additive" is the way how to make. As the title of this workshop says, the key word is "predictable". This is how and why people love AM. It is predictable; hence people can trust it, Prof. Han said.

4.5   Aircraft Structure Technology of Addictive Manufacturing

Prof. Xiangming Wang began by introducing the background of the aircraft structure technology. An aircraft structure requires light weight, high efficiency, long lifetime, low cost, and rapid manufacturing. Although it has been over half a century since the jet engine aircraft was invented, the aircraft structure has not changed much. This is because the traditional manufacturing technologies, such as forging, casting, welding, milling, turning, and grinding, have not changed much. With these complicated technologies, manufacturing an aircraft has long been high-cost. Therefore, an innovative structure technology is needed. The birth of AM provides an opportunity. With AM, large integral structure, material graded structure, function integrative structure can be expected.

Obtaining a large integral structure was almost impossible in the past because it has to be manufactured by combining tens of thousands of components. However, if the parting planes in design are eliminated, the redundant structures will also be involved in load transfer, which is preferred. But how can such a huge structure be manufactured? Since the strength and fatigue properties cannot be guaranteed by traditional manufacturing technologies such as welding, AM may find a way. The next question is how to avoid or control the deformation and crack such that the performance is achieved, Prof. Wang said. Global discretization-partition optimization has been applied by Prof. Wang's research group to achieve forming and connection at the same time. They proposed an equivalent Kt method for evaluating the fatigue performance of a structure. In forming connection, the mechanical properties are controlled by the parameters of forming process and heat treatment. Prof. Wang's group proposed a set of testing steps to test the components at three levels: small, medium, and large, where the medium is especially important. For a real typical component, passing the tests is not easy.

By AM, the component is fabricated with dissimilar metal material. In this way, the gradient distribution of the mechanical properties is achieved, and the material graded structure is formed, Prof. Wang stressed. The advantages are obvious. For example, the material layout is designable, the lightening efficiency as well as fatigue life are increased, and the integration of load bearing, heat and corrosion resisting are achieved. Prof. Wang then listed three combinations of different materials, i.e., TC4 + TC11, TA2 + TA15, and 300 M + A100. He showed the performance of the transition zone, including the geometric feature, specimen, static behavior, and fatigue behavior. The fracture toughness was briefly introduced. An important issue for material graded structure is the design and assessment method. Prof. Wang introduced two approaches. The first one is to develop new elements, which was finished in collaboration with Beihang University of China. With new elements, the crack propagation life analysis was established. Another approach is the multi-material layout optimization, which was finished in collaboration with Dalian University of Technology of China.

The functional system and fuselage structure were designed separately in the traditional structure, and the functional system was built on the fuselage structure. As a result, the design has some redundancies, and the weight increase is inevitable. By contrast, the fusion design in the functional integrative structure yields simple structure as well as light weight. Prof. Wang showed some metal multi-functional lattice structures fabricated by AM, which are actually the millimeter scale trusses where the optimal design of the unit cell configuration is one of the most important issues. From an engineer's point of view, Prof. Wang cares about the structural features, functional features, and process characteristics. To carry out the multi-constraint optimization, some issues must be solved, including load bearing characteristic, failure mechanism, failure mode, failure rule, and failure criterion. Some experiments and testing were done by Prof. Wang's group. However, this is just the beginning, Prof. Wang added. When analyzing the failure mode of the metal multi-functional lattice structure, some results were showed. For example, under compression, the length of broken rods is approximately equal to the unit cell rods, i.e., break occurs at the nodes. The reason is that the nodes yield first due to the highest stresses occurred at the nodes, and break occurs at the nodes after the buckling of the rods. According to Prof. Wang, the factors affecting the load bearing capacity of a metal multi-functional lattice structure include the unit cell configuration, the relative density of the structure, and the load types. The failure modes can be roughly categorized as plastic buckling, brittle fracture, connection fracture, and the buckling based wrinkling of the panel. Five steps were observed in the compressive test, i.e. linear elastic, soften, contact, compaction, and densification. It is also found that the relative density of lattice truss cores sandwich structure has a great impact on the compressive strength. Another test, the tensile test, was conducted, which reveals that the load-displacement curve is very similar to that of the traditional metallic materials. The observation in three-point bending test indicates that the top panel yields before the bottom one does.

Prof. Wang then briefly showed some engineering applications, including four crafts, eight materials, ten components, and a few model applications. He said that AM is useful in solving three problems: the bottlenecks of structure design, rapid trial manufacturing, and innovation by creative idea.

Prof. Wang pointed out some research areas in the end of his presentation, including developing new concept structures, new failure modes and criterions, new modeling and analysis methods, multi-constraint optimization involving technological properties and functional constraints, simulation techniques of forming process, integral assessment methods for structures, and testing methods of functional integrative structure.

4.6 Discussion

Regarding to Prof. Xu Guo's talk, the question was posed by Prof. Wing-Kam Liu from Northwest University. He noted that the approaches proposed in the presentation only focus on the geometric optimization. During the optimization process, the material property must be provided. When you do not know the material property of the design, how to consider it in the formulation from the material aspect? Prof. Xu Guo answered that only linear elasticity problem is considered in his approaches. Therefore, optimizing materials and material properties simultaneously is still a challenging problem.

Regarding to Prof. Michael Yu Wang's talk, the first question was proposed by Prof. Gengdong Cheng, he said he was interested in the problem Prof. Wang addressed, then he gave his comments about the homogeneous microstructure and uniform microstructure that it has been previously documented that uniform microstructure is not the best for topology optimization. The referred paper is (Liu, Yan, and Cheng, 2008). More works can be done about this topic.

Following Prof. Pinlian Han's presentation, there is a discussion moderated by Prof. Huiling Duan from Peking University. A participant asked about the meaning of concurrent AM. Prof. Han responded that concurrent AM is a way to parallel go on to hand all different options in AM. A question was asked about how much small can AM achieve for a turbine blade. Prof. Han stated that quite a lot of technologies associated with the requirements are restricted by manufacturing. With AM, one can make the entire part as needed. Making small pieces is possible. Another participant commented that combining one function with other functions to design a jet engine is important. If one wants to minimize the damage induced by birds strike on the blade, the traditional design based on stiffness is not good because energy absorption for that type of structure is bad. However, for foam-like material, the energy absorption is very good. Combining the two types of materials, one can create a new material which is load supporting and energy absorptive. Designing such material might be a challenge, the participant said.

Following Prof. Xiangming Wang's presentation, there is a discussion moderated by Prof. Jimin Chen from Beijing University of Technology, China. A participant commented that the theory does not catch up with the engineering because the structures manufactured by AM have been used in the aircraft by Prof. Wang et al. The same participant suggested that the properties of the structure have been changed by AM, which are very different from the foam metal structure. Another participant commented that addressing the fatigue problem should be focused in the AM based design.

Chapter 5
Additive Manufacturing Experimental Methods and Results, and Additive Manufacturing Scalability

The fourth session of the first two days of the workshop focused on the experimental studies of additive manufacturing (AM) and some applications in aerospace, medicine and so on. Prof. Lyle E. Levine (National Institute of Standards and Technology), Prof. Anthony Rollett (Carnegie Mellon University), Prof. Dongjiang Wu (Dalian University of Technology), Dr. Lianfeng Wang (Shanghai Aerospace Equipment Manufacture), Prof. Jimin Chen (Beijing University of Technology), Prof. Yonggang Zheng (Dalian University of Technology) respectively, discussed research results, challenges and future directions related to this topic.

5.1 Additive Manufacturing of Metals: An Integrated Materials Approach

Prof. Lyle E. Levine started the talk with a research question: how does the manufacturer convince the customer and regulatory agencies that a given part will meet specifications? Prof. Levine pointed out that a critical factor in AM of metals is the part qualification. Since AM produces the material and part geometry simultaneously, experimentally-validated simulations will be needed to predict the location dependent material microstructure and mechanical behavior. Prof. Levine outlined that the National Institute of Standards and Technology (NIST) are addressing this need through development of (1) quantitative in situ thermography for measuring the part temperatures during the build process, (2) thermography-validated thermo-mechanical finite element modeling to predict residual stresses and the internal, location-specific thermal history, (3) thermo-kinetic modeling (CALPHAD-based and phase field) of location-specific microstructural evolution using the modeled thermal history as an input, (4) extensive quantitative validation measurements (synchrotron X-rays, neutrons, electron microscopies, lab X-rays, quantitative metallography, dilatometry,

© The Author(s) 2018   53
X. Guo et al., *Report of the Workshop Predictive Theoretical, Computational and Experimental Approaches for Additive Manufacturing (WAM 2016)*, SpringerBriefs in Applied Sciences and Technology, https://doi.org/10.1007/978-3-319-63670-2_5

mechanical testing, etc.) at every step, and (5) an AM metrology testbed for developing new quantitative in situ measurements and producing highly controlled test builds.

Prof. Levine stated that NIST makes extensive use of large national measurement facilities for validation testing, which include the BT8 residual stress instrument at the NIST Center for Neutron Research (NCNR), the 34ID-E microbeam diffraction instrument (Levine et al. 2015) at the Advanced Photon Source (APS) at Argonne National Laboratory, and the 9-ID-C ultra-small angle X-ray scattering instrument (Zhang et al. 2016) at the APS. NIST is a partner on both of the APS instruments. The NCNR and APS microbeam diffraction instrument allow location-specific macroscopic and microscopic (sub-micrometer) measurement of full stress tensors from AM samples. The USAXS instrument uses simultaneous USAXS, pinhole SAXS and X-ray diffraction (XRD) to characterize microstructures from the atomic scale up to 10's of micrometers, both ex situ and during in situ heat treatments.

Prof. Levine pointed out a good example of the usefulness of these combined measurement and modeling methodologies is found in a recent publication by (Idell et al. 2016), where synchrotron USAXS/SAXS/XRD measurements were conducted in situ during a conventional 1066 °C homogenization heat treatment of an AM nickel-based super alloy, 718plus (Fig. 5.1). Unexpectedly, the deleterious, delta phase developed and subsequent thermo-kinetic modeling identified the AM, produced compositional gradients as the cause. Prof. Levine commented that such combined measurement and modeling studies are critical for improving AM part qualification and form the basis for their recent work on benchmark test development.

Prof. Levine pointed out that NIST are also reaching out to the broader international community by spearheading the development of a continuing series of AM benchmark tests (AM-Bench), with a corresponding conference series, which will allow modelers to test their simulations against rigorous, highly controlled AM benchmark test data, and will encourage additive manufacturing practitioners to develop novel mitigation strategies for challenging build scenarios. Prof. Levine described that at the moment, 61 members from 42 organizations are acting in the exploratory committee and their roles include identifying customers, defining classes of needed benchmark tests with example ideas and AM-Bench organization structure. He suggested that the next steps will be recruiting international steering committee, assembling NIST team for the first round benchmark tests and conference and recruiting external collaborators from numerous organizations. At the end of his presentation, he outlined the AM research timeline: for the short term (<3 years), the goals include (1) multiple open-platform testbed machines become available, (2) custom alloys receiving increased attention, (3) first cycle of AM-Bench tests completed, first conference held, (4) increased work on multi-faceted topology optimization; for the medium term (3–5 years), the goals include (1) major improvements in high fidelity models, integrated validated platform for AM simulation delivered, (2) commercial reduced order model simulations for

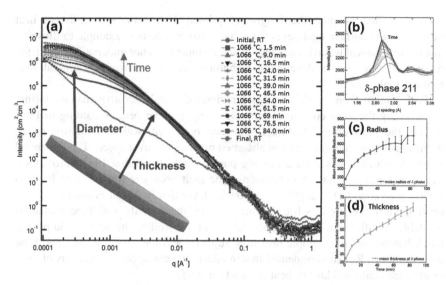

Fig. 5.1 (a) In situ USAXS/SAXS data acquired during an isothermal anneal of AM 718plus at 1066 °C along with the (b) in situ evolution of the δ-phase (211) peak as a function of time. Extracted time evolution of the δ-phase precipitates' (c) mean radius and (d) mean thickness

'design forward' approach being evaluated, (3) compositionally graded manufacturing becomes increasingly mainstream research; and the long term (>10 years) goals include (1) quantitative process monitoring become available, (2) some compositionally graded materials being used, (3) several measurement standards, testing standards, and best practice guides for AM in place.

5.2   3D Characterization of Additively Manufactured Materials, Including Synchrotron–Based X-Rays

Anthony Rollett recognized contributions from the following researchers: Jon Almer, Edward Cao, Ross Cunningham, Peter Kenesei, David Menasche, Tugce Ozturk, Suraj Rao, Hemant Sha, Samikshya Subedi, Chasen Ranger, Paul Chao, Jack Beuth, Elizabeth Holm, Fred Higgs, Tao Sun, Cang Zhao, Robert Suter, and Xianghui Xiao.

Prof. Anthony Rollett began by providing a global view from Terry Wohlers' 2016 annual report on additive manufacturing to show that metal powder bed systems were increasingly popular and most AM machines came from outside the U.S. despite that the U.S. is the biggest user of AM. To promote the research on AM, Carnegie Mellon University (CMU) has launched the Next Manufacturing Center with member companies including Alcoa/Arconic, Ansys, Bechtel Marine Lab., Bosch, Carpenter, General Electric, General Motors, Ingersoll Rand, FAA, Natl. Energy Techn. Lab., SAE International and U.S. Steel.

Prof. Rollett adopted the fatigue of printed Ti-6Al-4V with different heat treatments and surface treatments (Li et al. 2016) as a context example to illustrate that the mechanical behavior of these materials must be better understood before the benefits of this rapidly developing technology can be utilized for critical load bearing applications.

Prof. Rollett then provided a brief review of metals additive manufacturing, emphasizing the most common technology, viz., powder bed. According to Prof. Rollett, the most critical feature is the melt pool size, which is, to first order, determined by a trade-off between absorbed power and travel speed. The Rosenthal equation, which is based on a moving point source (Rosenthal 1941), provides a useful first estimate for the dimensions. The melt pool size must in turn be controlled in relation to the depth of the powder layer that is spread in each pass. The cooling rates are high, $\sim 106/s$, which means that rapidly solidified metals is available in bulk form. However, according to Prof. Rollett, the actual weld pool is much less regular than simple heat flow suggests and the variability has had little attention. Prof. Rollett commented that in effect, one can apply the analysis of laser or electron beam welding in heat conduction mode.

Prof. Rollett stressed that it is important to understand the microstructure and, in particular, porosity in additively manufactured metallic parts. Absent manufacturing defects, pores are the primary origin of fatigue failures under cyclic loading, for example. The morphology and location of these pores can help indicate their cause; lack of fusion pores with irregular shapes can usually be linked to incorrect processing parameters (Tang and Pistorius 2016), while spherical pores suggest trapped gas (Cunningham et al. 2016).

Prof. Rollett presented a model to calculate the porosity and density of additively manufactured materials. Prediction of porosity is a complex problem because the geometry of melt pools is complex; the pools overlap across layers, and there can be regions of unmelted material resulting in lack of fusion porosity. However, models are being developed that appear to match well with experimental data.

Prof. Rollett illustrated the characteristics of powder particles by showing the powder size distributions for EOM and Arcam systems. The powder sizes were determined on individual particles in SEM/optical micrographs. He stated that gas-atomized powders generally display a log-normal size distribution (Neikov et al. 2009) and the log-normal distribution will appear linear on adjusted cumulative probability plot. He commented that deviation from log-normal suggests sudden change in distribution (sieving). However, according to him, AlSi10Mg powder does not deviate from log-normal and EOS Ti-6Al-4V does not follow this trend.

Prof. Rollett stated that Synchrotron-based 3D X-ray microtomography was performed at the Advanced Photon Source (APS) on additively manufactured samples of Ti-6Al-4V using electron beam powder bed and Al-10Si-1 Mg using laser powder bed (Rao et al. 2016). He commented that the synchrotron source is useful for computed tomography due to its advantages in sample size, resolution and scan times. The spatial and size distributions of the porosity over a range of processing conditions were determined. Marked variations in the type and amount of porosity were observed as a function of the melt pool area. Outside of incomplete melting and keyholing, porosity appears to be inherited from pores or bubbles in the powder. Rollett noted that evidence for this inheritance is provided by preliminary results of high speed radiography at the APS that reveal the motions of individual particles in miniature powder beds under the action of a laser pulse. He stressed that not only can the melt pool be imaged along with particles flying off but also bubbles within particles are visible as they merge into the melt pool.

Prof. Rollett commented that beyond measurements of porosity, 3-D printed parts are known to have residual stress as a consequence of the shrinkage that occurs on solidification as well thermal contraction. According to Rollett, thanks to recent advances in high-energy (synchrotron) X-ray methods, a combination of near-field and far-field high energy diffraction microscopy (HEDM) enables the mapping of both 3-D grain structure and the lattice strains (Schuren et al. 2015). Preliminary measurement results are presented for printed Ti-6Al-4V. Remarkably enough, both the majority hexagonal phase and the minority bcc phase can be reconstructed. Moreover, parent bcc orientations inferred from the product hcp material agree well with the HEDM reconstructions of the bcc grains. Prof. Rollett stated that once such data are available, the impact of microstructure on properties can then be evaluated.

Prof. Rollett gave a description about the application of image-based spectral methods (e.g., Michel et al. 1999) for calculating the micro-mechanical response, where the measured image is used as direct input. The results show that the degree of mechanical anisotropy between the build direction and orthogonal directions is almost negligible in Ti-6Al-4V despite the strongly columnar microstructure of the beta phase as deposited. Prof. Rollett stated that this can be rationalized in terms of the multiple orientation variants observed as the alpha phase forms in the beta such that the texture of the majority phase (alpha) is nearly random.

Lastly, Prof. Rollett summarized his speech by emphasizing that understanding microstructure is important during every step in the additive manufacturing process. He mentioned that CMU in collaboration with APS et al., is addressing this challenge by combining

- Measurement of powders and defects, especially pores: analysis with extreme value statistics, link to powder flow.
- Advanced characterization 3-D microscopy with high energy synchrotron X-rays, e.g., tomography of voids.
- Demonstrated ability to predict incomplete melting; links to studies by AFRL of porosity varying with location in the build volume.
- Machine vision is able to classify powders using images, going beyond even the human expert.
- Development of micro-mechanical models for materials used in AM: negligible anisotropy seen in modeling agrees with experiment.
- Advanced heat exchangers, dimensional limits.

Rollett raised the following open questions, which are important to the further development of AM, including:

- Source of porosity, excluding incomplete melting and keyholing (excessive power density); e.g., gas solubility (Elmer et al. 2015).
- Effectiveness of hot isostatic pressing? Evidence that pores re-open.
- Predictability of weld/melt pool variability; radiation pressure, Marangoni effect, behavior of bubbles in melts, gas trapped during powder melting.
- Control of solidification microstructure.
- Absorption by powders of light and interaction with e- - beams.
- Constitutive relations (EOS, shear strength) for AM materials.
- Phase transformations in titanium (and other metals).
- What characteristics of powders change with recycling?
- Trade-offs between powder size (cost), surface finish, build rate etc.
- Optimization of alloy composition against required properties.
- How best to apply data analytics, and how to share data.
- Machine to machine variability.
- Predictability of residual stress: at the part scale and at the grain scale.
- Integration of part design with build design.

5.3 Direct Additive Manufacturing of Ceramics by Laser Engineered Net Shaping

Prof. Dongjiang Wu introduced the direct additive manufacturing of ceramics (DAMC) conducted by laser engineered net shaping system with pure ceramic powders. Via this approach, pure ceramic powders without binders are directly melted by laser beam and solidified to form the designed structures layer by layer. This method combines the conventional fabrication processes of mixing, shaping, drying, and sintering of the ceramic into a single procedure. In his study, Al_2O_3 was used as a building block to study the feasibility of DAMC, as well as the mechanism and influencing factors of the crack propagation during fabricating process. A cracking criterion and a series of crack suppressing methods were proposed, such as reducing coefficient of thermal expansion, particles toughening and fine grain toughening. Crack-free structures of hollow blade, cone and cylinder with the dimension of 30 mm were successfully deposited by Al_2O_3, Al_2O_3/ZrO_2, Al_2O_3/TiO_2 and $Al_2O_3/TiO_2/SiC$. Ceramics with dense microstructures generated by the melting-solidification process possesses the micro-hardness over 2100 HV and the fracture toughness over 4.79 $MPa\cdot m^{1/2}$. These mechanical properties exceed conventional hot-pressed sintering level, indicating that DAMC is a promising technique for the production of high-performance ceramic structures in a single step, he pointed out.

Prof. Wu began by introducing ceramics, an important inorganic non-metal material, which has been widely utilized in many areas, such as mechanical, electrical and information engineering as well as the biomedical engineering due to its outstanding properties. However, the impurities and porosities with conventional Mixing-Shaping-Sintering manufacturing methods limit the potential industrial applications. Conventional process for fabricating ceramic components is Powder Preparing-Shaping-Drying-Sintering. Doping of binders or plasticizer and its solid-phase sintering mechanism decide that fabrication of dense ceramic components with high-purity and complex geometric is time-consuming and hard to achieve, he explained. In contrast, DAMC is a subversive process method for fabricating ceramic components, by which pure ceramic powders without any binders or plasticizers can be directly and completely melted by high-energy source and then deposited to form net-shaped structures layer by layer. He also pointed out that because of the melting-solidification process, the purer and denser ceramics and even functional components could be obtained directly. However, cracking is a hard problem in DAMC because of high thermal gradient during deposition and intrinsic brittleness of ceramic material, he noted. Hence, he has done much analytical and experimental work to suppress cracks in DAMC.

In their research, laser engineered net shaping (LENS) system was used to study the feasibility of direct fabrication of Al_3O_2-based ceramic structure. Their LENS system combined the YAG Laser, three powder feeders and coaxial nozzle into CNC machine. Prof. Wu pointed out two experimental conditions in his report. First is that the ceramic powders used in their experiments is high purity (>99.7%), with no binders or plasticizer and only need drying before fabrication. Another is that the substrate material would have a good physical compatibility with the sample material.

Prof. Wu stated that single factor experiment and orthogonal experiment methods are not particularly suitable and effective for fabricating ceramics. This is because that ceramic material does not have good plasticity like metal. Hence, he proposed a set of process model to guide the process optimization in DAMC. By simplifying the morphology of the top thin-walled sample structure and analyzing the deposition process, he found that the forming process of DAMC results in a conservation of mass relation. According to the mass conservation and energy conservation relationships, fabrication process parameters of required specific size sample can be calculated (Niu et al. 2014). All the reasonable combination of process parameters can be got by the process model, he explained.

Cracks are the primary issue in DAMC, the cracking mechanism is related with the material properties and process conditions. A good understanding of cracking mechanism in DAMC will be significant for suppressing cracks. According to theory of thermal elasticity and fracture mechanics, Wu analytically expressed the thermal stress during deposition and fracture strength of specimen, and proposed a cracking criterion. When the thermal stress exceeds the limit of fracture strength, the initial cracks will extend. The criterion clearly reveals the influence factors for cracking propagation (Niu et al. 2016), including thermal expansion coefficient, fracture surface energy, initial crack strength and so on.

Based on the cracking criterion, Prof. Wu also proposed a series of crack suppression methods. Increasing the fracture strength of material and reducing the thermal stress are two important directions. Because of the lower thermal expansion coefficient of aluminum titanate than alumina, one of these methods is combining titanium oxide into alumina to generate aluminum titanate. The principle is to lower the thermal expansion coefficient, thereby reduce the thermal stress. Many different structure samples, as shown in Fig. 5.2, are fabricated successfully with low thermal expansion coefficient material Al_2O_3/TiO_2. The micro image showed that aluminum titanate forms a continuous net shape which separates the grain of alumina. The flexural strength of these samples is over 200 MPa and micro-hardness is about 21 GPa. Hence, Prof. Wu pointed out that this material is promising in the field of biological implantations.

Another method he proposed is particle-reinforced, by doping silicon carbide into Al_2O_3/TiO_2, because silicon carbide has a higher fracture surface energy than alumina. The principle is to enhance the fracture surface energy of the material, thereby increasing the fracture strength. A cylindrical structure sample with length of 150 mm is fabricated successfully with $Al_2O_3/TiO_2/SiC$. According to the micro images of sample, pinning, trans-granular crack and crack deflection are observed when crack meet silicon carbide particles. Silicon carbide can also eliminate porosity during the alumina and titania reaction.

He also proposed a method called grain refining to suppress cracking. This was achieved by mixing zirconia into alumina, which can produce eutectic reaction and generate fine eutectic structure. The eutectic microstructure graphs of Al_2O_3/ZrO_2 showed that the fine flake eutectic parallel to the deposition direction and the eutectic spacing is about 100 nm, which is much smaller than that of pure alumina. The principle of this method is decreasing the initial crack strength to increase the fracture strength. Many different eutectic microstructure samples were successfully fabricated with almost no crack, as shown in Fig. 5.2. He also examined the properties of this material. The results showed that the relative density of the material is 98.3%, micro-hardness is 17 GPa and fracture toughness is 4.79 MPa·m$^{1/2}$. Doping yttria into alumina can achieve the eutectic microstructure

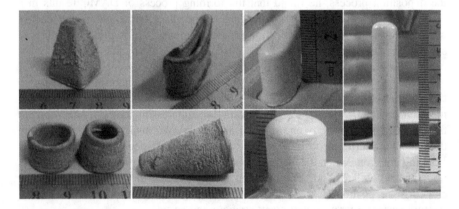

Fig. 5.2 Ceramic structures fabricated by laser engineered net shaping

of Al_2O_3/YAG. The eutectic spacing is about 300 nm and the relative density of the material is 98.6%. The micro-hardness of the sample fabricated with Al_2O_3/YAG is 17.35 GPa and fracture toughness is 3.14 MPa·m$^{1/2}$.

Functionally graded structures are also fabricated successfully using different mixing ratios of Al_2O_3/ZrO_2. It's only the beginning of the future work, he explained.

In conclusion, Prof. Wu summarized five conclusions from his presentation: (1) Direct additive manufacturing of high-performance ceramics without binders can be achieved by Laser Engineered Net Shaping; (2) Cracks is the primary issue in DAMC, and the cracking mechanism is related with the material properties and process conditions; (3) Al_2O_3/TiO_2, Al_2O_3/ZrO_2, Al_2O_3/YAG were successfully fabricated by DAMC and demonstrated high density and properties; (4) Functional graded eutectic ceramic also can be achieved by DAMC; (5) Increasing the strength of material and reducing the thermal stress are two important directions for fabricating components with larger size.

Finally, he highlighted some future researches on DAMC. Short term researches include: (1) Meet property requirement of actual part; (2) Achievable fabrication size; (3) Achievable types of processing materials. Intermediate term research is how to quantitative control of property and dimension accurately. Long term researches contain: (1) Optimization of cycle time and cost; (2) Application and promotion of industrialization.

5.4 The Exploitation and Application of Data-Driving Model in the Mechanical Property Prediction of the Critical Addictive Manufactured Space Components

Lianfeng Wang recognized contributions from the following researchers: G. Meng, L.J Guo, and P. Li.

Dr. Lianfeng Wang and his research team focused on the investigation and development of the 3D printing technology to design and produce the space components several years ago, and they paid attention on the application of 3D printing and the relevant products but not the underlying fundamental theories. He began his talk by presenting a picture of their self-fabricated equipment for AM, and he introduced their own understanding on the design concept during the fabricating process of this equipment, including the free design for industry, the advanced design through hybrid manufacturing and the potential application in rapid repair of defective parts.

Dr. Wang commented that the 3D printing could overcome many difficulties, such as the form complexity, the material complexity, the layer complexity (multi-layer structure) and the function complexity, which are unachievable by the traditional manufacturing approaches. It can provide a new and accessible method for the designers to "create without limits" under the concept of "what you think is what you get". In the aerospace engineering, the integral manufacturing based on the 3D printing could improve the safety and reliability of products by reducing the component number. In the future, 3D printing will play an important role in the design and fabrication of new generation of aerospace devices with high-speed, high mobility and high precision.

Afterwards, Dr. Wang showed some current applications of the AM technology, such as the satellite, the aircraft, the propulsion, the international space station, the rocket engine, the aerospace engine impeller, the unmanned aerial vehicle, the radar shell, etc. In these applications, many complex or functional components, which are not possible to be produced by the traditional manufacturing approaches, could be fabricated by AM with a short design period and a low cost. However, the level of the present AM technology still limits its further application to the critical components of the aerospace devices.

The properties of materials fabricated by AM could be influenced by many factors. Here, he showed some pictures of the microstructures and tensile deformations of 3D printed specimens based on the selective laser melting technology to examine the effects of the scanning strategy (chessboard and meander strategy at the preheating temperatures of 80 and 120°C, respectively), the style of heat treatment (on the top and side surfaces), the inclination angle and the support parameter.

In the end of his presentation, Dr. Wang pointed out that AM in aerospace will be developed towards the fabrications of large-size components and fine structures. Nowadays, the velocity and precision of 3D printing still cannot meet the requirements of industry. They will look forward to and keep paying attention on the new technologies in AM, such as the high-precision printing, hybrid manufacturing, etc. Moreover, he talked about his understanding on the further development and application of AM in aerospace, including:

- The fabrication of large-size and complex components of the rocket and satellite;
- The simple maintenance of aerospace devices through rapidly fabricating specific tools or components in space;
- The fabrication of aerospace devices by using the local resources in the space or on the planet.

5.5   The 3D Printing Technology for Medical Health

Jimin Chen recognized contributions from the following researcher: Yansheng Li.

Prof. Jimin Chen began by noting that his presentation would focus on the application of 3D printing technology on the medical field. He introduced some products of their 3D printing center, including the acetabular cup and customized shoe pad.

Firstly, Prof. Chen reviewed the development of the 3D printing technology. He introduced some representative printing methods such as the stereo lithography appearance (SLA) and digital light processing (DLP) based on liquid, selective laser sintering (SLS) based on powder, fused deposition modeling (FDM) based on wire, etc. Then, he summarized the advantages and disadvantages of 3D printing technology and traditional manufacturing technology.

The second issue he raised is that the application of 3D printing on medical health. He commented that the digital medical 3D printing technology is a new technology which combines traditional medical treatment with digital design, computation, information, biological engineering, and material science. It is closely related to people's livelihood and may bring a great change for people's lives, health care and traditional business model. With the development of technology, the applications of 3D printing in the medical field will increase rapidly (Jonathan and Karim 1997; Madamesila et al. 2016). Basically, there are four levels of the applications of 3D printing in medical service, from the outside of human body into the body, from simple to complex cases, which can be described as:

- 3D printing models;
- 3D printing tools;
- 3D printing implant;
- 3D printing for tissue engineering and human organ.

The third issue is the application of 3D printing in tumor surgery. Firstly, he introduced the tumor treatment methods such as surgery, radiotherapy, chemotherapy, biological therapy, traditional Chinese medicine or combined treatment. Under a high temperature or through some radioactive methods, some chemical drugs can kill cancer cells, but the biggest problem is how to separate the normal tissues and cancer cells, he stated. The therapy of radio particles implant was proposed. He said that the idea is to put the particle into the tumor to prevent cancer cells from radiating, which is commonly referred to as radioactive particles implantation technology. The particles are radioactive iodine 125 with the diameter of 0.8 mm and length of 4.5 mm. One little particle can kill cancer cells of about 4–5 cubic centimeters. For one tumor patient, according to tumor's size, it usually needs 30–40 particles to kill all cancer cells. Although this method has many advantages, the actual application and successful cases are rather limited. The main reason is how to implant so many tiny radioactive particles uniformly into the entire tumor. In most cases, the surgery usually determines the particles position under the guidance of ultrasound or X-ray. He explained that the whole system is complex and not easy to operate, and it is difficult to determine the position accurately so that the curative effect is not satisfactory. However, 3D printing technology can improve this situation. He designed and calculated the number of implantation of radioactive particles that could completely kill cancer cells according to tumor shape and size. He followed the steps: the first step is to obtain the tumor size and shape information, which is used to reconstruct the model of tumor in the computer for the calculation of the required quantity and arrangement of particles; the second step is to put the particles into the tumor in vivo; the last step is to put the particles in the right position. He emphasized that the whole process can be simulated in advance in

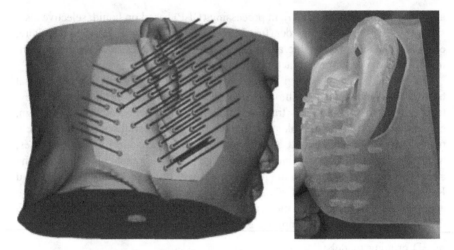

Fig. 5.3 Simulated and printed guiding plate by 3D printing technology. *Source* Prof. Jimin Chen, Beijing University of Technology, presentation to the workshop

computer before operation. In order to finish the desired operation, 3D printing technology has to be applied. Doctors can design and print different plates by 3D printing technology for different lesions. In surgery, doctors just put the pinhole plate on the lesions and insert needles one by one, he said. Figure 5.3 shows the simulated and printed guiding plate by 3D printing technology. Finally, he gave some clinical examples of 3D printing in tumor surgery.

In conclusion, Dr. Chen commented that the 3D printed guiding plate makes the surgery effective, and he stated that the 3D printing technology possesses big market in medical area. Then, he suggested that 3D printed medical parts should be approved by CFDA. At last, he emphasized that talent people are urgently required in this crossing field and the cooperation is necessary for the application of 3D printing technology in medical health.

5.6 Recent Simulation Studies on the Sintering, Grain Growth and Coupled Thermo-Mechanical Process of Metallic Materials

Yonggang Zheng recognized contributions from the following researchers: Zhen Chen (Dalian University of Technology & University of Missouri), Jun Tao (Dalian University of Technology & University of Missouri), Cen Chen (Dalian University of Technology), Hongfei Ye (Dalian University of Technology), Jiayong Zhang (Dalian University of Technology), Peiqiang Guo (Dalian University of Technology), Hongwu Zhang (Dalian University of Technology).

Prof. Yonggang Zheng gave a brief introduction about the collaborative efforts in the modeling and simulation related to the multiscale problems in powder based additive manufacturing (PBAM) process at the International Research Center for Computational Mechanics of Dalian University of Technology, mainly including the atomistic understanding of laser sintering of nanoparticles, the developments of the lattice Boltzmann model for the simulation of grain growth and the generalized interpolation material point method for the analysis of coupled thermo-mechanical process.

He pointed out that tuning the material and structural properties based on numerical simulations in the AM processes is of great importance for practice. Currently a large number of AM processes are available and they are mainly different from each other in operating principle, depositing way and materials used (Bikas et al. 2016). Among these processes, the PBAM processes, including the SLS, selective laser melting (SLM) and electron beam melting (EBM) processes, become particularly attractive due to their inherent advantage to the AM of load-bearing structures from metal, which have great potential applications in many areas, including the aerospace engineering.

Prof. Zheng further reviewed the multiscale problems in the PBAM of metallic materials, as shown in Fig. 5.4. He emphasized that, since the SLS, SLM and EBM processes melt materials to produce components and structures, the related AM processes usually involve melting and solidification of materials, heat transfer, grain growth, erosion, and so on. The performances of the products are also closely related to the heat source used. The products may present different surface roughness, residual stresses and many other properties. Moreover, these behaviors

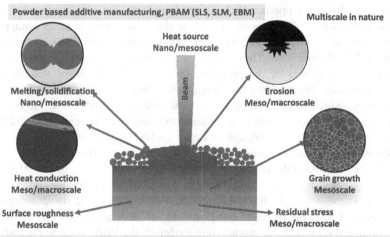

Fig. 5.4 Multiscale nature in additive manufacturing of metallic components based on the selective laser sintering, selective laser melting and electron beam melting techniques. *Source* Prof. Yonggang Zheng, Dalian University of Technology, presentation to the workshop

usually occur or dominate at different scales. Such as, the melting and solidification usually occur at nanoscale and mesoscale, the grain growth occurs at mesoscale, while the overall heat transfer occurs at macroscale. The spatial and temporal scales may span from nanometers to meters and from picoseconds to hours, respectively. There are various challenges in related numerical simulations due to the multiscale nature of the AM of metallic materials and structures.

Although much work has been done previously, as Prof. Zheng mentioned, it is still desirable to develop effective models and algorithms at different scales and a multiscale modeling platform for the PBAM is highly demanded. In their work, three fundamental issues in the modeling and simulation of the PBAM of metallic materials have been addressed to advance the understanding of structure-property relationship at various scales related to AM, including the mechanisms for the melting-solidification processes, the computational algorithm for grain growth behaviors and the numerical method for thermo-mechanical behaviors.

He first talked about numerical understanding of the sintering behavior of nanoparticles. In these studies, the molecular dynamics method showed great potentials to reveal the atomistic nature and underlying physics of the nanoparticle sintering. The melting and solidification during the sintering of metal nanoparticle usually involve many different phenomena, such as the solid state sintering at room temperature, the melting and fusion of nanoparticles as the temperature increases and the solidification and crystal growth in the cooling stage. Some peculiar microstructures (such as the dumbbell-like neck and five-fold deformation twin) in the sintering process have been identified. The influences of the heating/cooling rate, and the vibration amplitude and frequency on the sintering behaviors have been explored.

The mechanical properties of the sintered products have also been explored. The final products are still sphere-like and their plastic deformation under compressive loads is mainly accommodated by grain boundary and partial dislocation activities due to their limited size. They found that the heating rate can greatly alter the mechanical properties of the products. Moreover, they observed that there is a correspondence between the maximum stress and the environmental vibration frequency, which is almost independent with the nanoparticle size. He explained that the results are important to understand the transition between solid and liquid, and the established procedure can be easily applied to many other materials or the sintering of a portion of metallic particles with a size of several microns.

He then discussed the numerical models related to the grain growth. Although a variety of numerical models have been proposed to model the grain growth behaviors, these models have their inherent advantages and disadvantages. He pointed out the merits and disadvantages of several representative models. Especially, it is not easy to extend these models to the PBAM due to the combined effect of high temperature gradient, transient heat conduction and melt convection. In their work, they proposed lattice Boltzmann models for the grain growth simulations, which can be further integrated with the lattice Boltzmann models for melting and heat transfer, thus show good promises for AM (Zheng et al. 2016).

Actually, he also demonstrated that the developed model can recover the conventional phase field model based on a multiscale expansion analysis and thus the model parameters can be readily obtained from various databases available. He illustrated the performance of the models by considering the shrinkage of a circular grain embedded in a matrix, the grain growth in a polycrystalline system, the grain growth in two-phase systems and the grain growth in various systems involving the temperature gradient.

Finally, Prof. Zheng explained that, despite the atomistic nature of sintering and the grain growth behaviors, the coupled thermo-mechanical process at mesoscale and macroscale is also an important aspect in AM processes that needs to be well studied. Moreover, failure may occur due to local heating. It is of great importance to develop effective numerical methods to simulate these behaviors. Thus, he and his collaborators established a generalized interpolation material point method for the coupled thermo-mechanical process (Tao et al. 2016). Particularly, the material points can be conveniently added or deleted during the computation, thus it is quite suitable to model the thermo-mechanical behaviors as presented in the layer-by-layer additive manufacturing. Various applications show that the proposed procedure for thermo-mechanical analysis may provide a valid spatial-discretization tool for multi-physics simulation.

Most importantly, high energy heat sources are usually applied to the materials in PBAM, it is of great importance that the numerical methods can effectively model the failure behavior due to local laser heating. In their work, a mechanical decohesion model formulated based on the discontinuous bifurcation analysis identifying the transition from continuous to discontinuous failure modes and a coupled thermo-mechanical decohesion model have been adopted and implemented into the generalized interpolation material point method. Through the simulation of a pre-stressed square plate subjected to a laser heating, it has been demonstrated that the decohesion failure due to localized heating can be well simulated by using the method.

In conclusion, Prof. Zheng emphasized that, although these developed models and algorithms have shown many promises in understanding the mechanics of AM processes at different scales, they, together with other methods, such as the widely used finite element method and the newly developed extended multiscale finite element method, should be integrated to model the powder based additive manufacturing processes. This is a long-term goal of the future work.

5.7   Discussion

After Prof. Lyle E. Levine's presentation, the first question was regarding the AM Benchmark Test Series and wondering whether they are open to everybody. Prof. Levine commented that for the AM Benchmark Test Series, he would like more people (academics, industrial people, etc.) from different organizations and different countries to get involved and then the next steps will be to assemble a steer committee, to define the structure, to work with the local committee to put ideas together and restart the benchmark test, and then to spread out the benchmark to the whole societies and communities. He also mentioned that if anybody who is interested in participation and discussion can send him an email and broad inputs are welcomed.

The second question was regarding the timeline and the participant was wondering whether there is an optimistic timeline for AM. Prof. Levine commented that at the moment the solid benchmark is really qualitative and can measure anything with no boundary and that will be made available so that people can verify their models using the benchmark test series. He especially mentioned that people from the software company who build simulation tool can use the benchmark verify their simulations against the real tests and there is huge opportunities for them to quantitatively validate their predictions. He closed his talk by saying over the next years there will be numerous development and it is not easy to put a timeline.

The first question concerning Prof. Anthony Rollett's talk was posed by a participant regarding the conclusion that anisotropy is negligible in the printed parts, because the participant wondered how the part manufactured layer by layer can perform as a uniform layer. The participant stated that, in contrast to the small size sample presented by Prof. Rollett, a larger size sample had been manufactured and anisotropy was indeed observed. Prof. Rollett explained that the conclusion is valid at least for the particular samples of Ti-6Al-4V in which the strongly oriented bcc phase initially formed transforms to hcp with many orientation variants. The phase transformation effectively randomizes the material and decreases the anisotropy. Other materials such as fcc metals that lack an allotropic phase change are more likely to exhibit anisotropy as-deposited.

A participant wondered whether the results would be affected if other types of power were used and he asked how to choose the power in AM. Prof. Rollett stated that he had demonstrated that, in fact, one can change the power by a large amount. The key is that users have to look at the layer depth and hatch spacing and must adjust the melt pool size by balancing the speed against the power. He commented that, providing that the users pay attention to the applicable process map, there would be no problem in changing the power.

The question related to Prof. Dongjiang Wu's talk was posed by a participant regarding if they doping the different powder materials in the first order during fabrication process. Prof. Wu explained that the powder materials are not mixed in advance. There are three cylinder powder feeders in the LENS system and the different powders in the different powder feeders during the fabrication process.

A participant posed a question regarding Dr. Lianfeng Wang's talk on how to enhance the fatigue strength of the printing materials induced by the existence of interfacial region. He commented that they have conducted some investigations on this issue but the relevant works are still on the way. The participant asked if the company has developed own software platforms or algorithms in this research. Dr. Wang answered that they only use the commercial software at the present stage, but the own software is indeed very significant, and they would carry out some relevant works in the future.

After Prof. Jimin Chen's talk, a participant posed a question regarding to the porosity and surface porosity of the materials during the 3D printing procedure. Prof. Chen responded that the key work is design, and the porosity is not a problem for the simple shape shown in the presentation. Another participant wondered if it is practicable for the traditional Chinese medicine such as acupuncture points and meridians by the 3D printing technology. He explained that the technology can be applied in any place as long as the picture of the human body can be captured. From this point, he thought the technology can be applied in the traditional Chinese medicine.

The first question concerning Prof. Yonggang Zheng's talk was posed by Prof. Wing-Kam Liu regarding the high heating rate used in the molecular dynamics simulation and how to identify the actual melting and solidification mechanism in AM. Prof. Zheng explained that the heating/cooling rate used in the molecular dynamics simulation is indeed very high as compared to that in the AM of metallic components for industrial applications. Their primary work mainly focused on the femtosecond laser sintering of nanoparticle due to the inherent limitation of atomistic simulation methods. How to expand the capacity of these methods or link these methods to other meso/macroscale methods is still an open research question.

The second question concerning Prof. Yonggang Zheng's talk was also posed by Prof. Wing-Kam Liu regarding why the lattice Boltzmann method has been chosen for the grain growth. Prof. Zheng explained that the current widely used Potts model, vertex model and phase field model for grain growth have their inherent advantages and disadvantages. The Potts model is simple but has some difficulties to scale the Monte Carlo time to the physical time. The vertex model is intuitive but very difficult to be extended to three dimensional systems. The phase field model seems to be the ideal one for grain growth simulation. However, recent advances on AM have demonstrated that the grain growth of materials during the selective laser melting may occur under high temperature gradient, transient heat conduction and melt convection. It is still impracticable to consider all these effects within the framework of phase field model. More recently, the lattice Boltzmann model has been proved to be a powerful candidate to model phase transition from solid to liquid then to solid in the PBAM. Thus, provided the lattice Boltzmann model for the grain growth simulations, a multiscale computational platform may be built up for PBAM.

Chapter 6
Summary of Group Discussions

During the third day of the workshop, participants got together to discuss the challenges and possible solutions to the questions raised in above sessions. The group discussions are summarized and provided in the following subsections.

6.1 Theoretical Understanding of Material Science and Mechanics in Additive Manufacturing

6.1.1 Topic Discussion Organizers

Prof. Lars-Erik Lindgren (Luleå University of Technology, Sweden) and Prof. Huiling Duan (Peking University, China).

6.1.2 Summary

The topic discussion organizers stated several key points in this topic on the theoretical understanding of material science and mechanics in additive manufacturing (AM). Several workshop participants then discussed the fundamental scientific issues of AM to be addressed, and emphasized the importance for the benchmark tests in material expression and model validation. They identified some of the key points to be further investigated in the field of computational mechanics:

- The quantitative measurement and validation in various processes and materials;
- The underlying physics of materials in AM;
- The link between the physics and chemistry and the traditional mechanics;
- Physical understanding of the microstructure;

© The Author(s) 2018
X. Guo et al., *Report of the Workshop Predictive Theoretical, Computational and Experimental Approaches for Additive Manufacturing (WAM 2016)*, SpringerBriefs in Applied Sciences and Technology, https://doi.org/10.1007/978-3-319-63670-2_6

- Efficient models for AM and calibration;
- Communication between the materials science and mechanics communities;
- Constitutive models for AM;
- The bridge between microscale and macroscale;
- Description of different material systems;
- Computational methods for the void formation and its related effects.

The group members discussed the fundamental points in AM and also addressed the importance for the establishment of Benchmark tests for model validation, the different model development strategies, and even the calibration. Some of them mentioned the further use of AM in composites.

The comparison between numerical predictions and measurements is the key for model selections and model developments. They mentioned how to select appropriate variables in measurements and how to compare them with the results from computational models. They also noticed the important points including the formulation of phase transformations and microstructural changes, theoretical developments for microstructural evolutions in different material systems, the link between micro scale variables and the macro scale properties in additively manufactured parts and the requirements of appropriate constitutive models in the numerical analyses.

They mentioned that the void formations can affect the success of AM and the subsequent fatigue properties in its service life. They also mentioned the size of the imperfection which needs to be determined in the online and offline monitoring. Some of them mentioned that the description of bubbles inside the melted pool and the heaping of small defects remain to be challenging.

6.2 Computational and Analytical Methods in Additive Manufacturing

6.2.1 Topic Discussion Organizers

Prof. Michele Chiumenti (Technical University of Catalonia, Spain) and Prof. Zhao Zhang (Dalian University of Technology, China).

6.2.2 Summary

The topic discussion organizers Prof. Michele Chiumenti and Prof. Zhao Zhang, collected and posted many questions regarding to the computational and analytical methods in AM, including

- Which are the available/best methods to bridge the spatial scales in AM?
- How do we best keep track of the material state (geometry, history variables …)?

- How can temporal scales be bridged?
- Which are alternatives to a detailed physical simulation?
- What role can data analytics/reduced order models/AI play?
- Which is the more suitable constitutive model?
- Is the down scaling to microstructure level feasible?
- Is defect prediction an issue for AM simulation tools?
- Is there more interest on global (structural) level simulation or local (melting pool)?
- How to bridge local and global level analysis?
- Which is more interesting to industry: accuracy or speed?
- Standardization: constitutive model from powder?
- Standardization: scanning sequence?

Prof. Chiumenti pointed out that the current computational technology is still far away from solving the problem (computer modeling of real AM process) and so far there is still not specific computer software for modeling AM. He stated in detail the multiple scales in both space and time in AM process and this poses a big difficulty for modeling AM. The multiscale and multiphysics nature of AM is emphasized. The constitutive models and the dependence of the parameters on temperature are also highlighted for AM modeling. The following questions are also discussed:

- How does a role model for customizing a process simulation software (algorithms) for a hardware supplier (material and process calibration) look like?
- How can we continue to define benchmark problems for the different aspects of AM simulation?
- What can you do that no unauthorized party 'hacks your structure'?
- Which AM-specific questions arise concerning IPR?
- Which issue is w.r.t. security and liability?

In order to obtain more accurate results of AM simulations, participants of the discussion session discussed the importance of bridging the spatial and temporal scales. The participants stressed that the selection of scale in computational model depends on what goal we want to achieve, as there is no model suitable for all scales, and we also need a quite simple approach for a specific problem. They also think that the model crossing scales should not only memorize the evolution of microstructures in history, but also show the current response to loads. The participants also discussed the importance of data analytics, reduced order model and artificial intelligence the Artificial Intelligence (AI). They think these approaches are useful for both engineering applications and scientific research to narrow down the scope of problems. Besides, the participants discussed the progress in defect prediction by using a scanning technique made by engineers in industry. They think that applying fundamental sciences into real problems is crucial to obtain progresses for researchers in this field.

Several participants discussed the role of the finite element method in AM. They pointed out that it is important to be aware that the simulation by the finite element method does not represent the exact consequence. There are time scale and spatial scale issues in the simulations. The subgroup members stated that the simulations at the particle scale are not fine enough to gain a complete understanding of the AM process, but that is the best the researchers can achieve at this time. They noted that it is the biggest computer's role to look at the individual particles to see how the particles can diffuse together and re-melt again. The time history is a very useful piece of information, but there is a much smaller scale than what the researchers can simulate nowadays. It is also pointed out by the subgroup members that there are a large number of other numerical techniques which are used by the researchers to try to gain much fine time scale and spatial scale in the simulations.

A number of participants emphasized the indispensable role of the computer codes in the simulation of AM. The subgroup member stated that the researchers in the field of computational mechanics have developed a huge number of codes, including the commercial codes, lab codes and individual codes. These program codes greatly help people in understanding a lot of different kinds of real-world processes. Taking the finite element method for example, various types of physical processes can be embedded into the final codes, as most commercial CAE software (such as ABAQUS and ANSYS) do. They stressed that it would be a giant task to integrate up these program codes, through which a complete map of the whole life cycle of AM can be obtained.

Many session participants noted that it is of great importance to make sure that the simulation models are correct in order to reflect the intrinsic characteristics of AM. It is highlighted by the subgroup members that before the researchers in the field of AM run their personal numerical models, they are required to know the exact processes, the mechanical behaviors of materials that they are going to carry on simulations. When commercial software is adopted in the simulation, the researchers are also suggested to understand the basic algorithms. The participants commented that the researchers should be aware of the assumptions which are made to build the models and judge whether these assumptions are reasonable. They noted that special attentions should be paid when the numerical methods based on the static equilibrium equation are adopted to simulate the dynamic process since it is a simplification of the original solution. The differences between the big models and small models were also addressed by the subgroup members.

Another topic discussed by the participants was the standardization of scanning sequence. A number of participants commented that it might not make sense to standardize the scanning sequence. The machine manufacturers basically have to adopt different standard sequences so as to avoid IP. They noted that this is a field where the AI and data analysis can play an important role. According to these subgroup members, what the researchers in AM have to cope with is about learning and getting information from the data analyst.

Several session participants commented that more effective cooperation between the researchers and the industry is needed to promote the development of AM in the future. The researchers in the field of AM sometimes fail to get what the industry really needs and become uncertain about whether the academic research would be useful in the industry. It was suggested that the industry could focus on practical AM and give expertise on what types of processes that need to be investigated in depth while the academic researchers could find out the underlying mechanisms of the observed phenomena and provide guidelines on what can or cannot be measured experimentally. It was also noted by the subgroup members that most of the industrial companies have their own computer codes, and the reason why they do not use the commercial code is because it is not available to them.

Several session participants also emphasized the need for academic researchers with different backgrounds to learn from each other, which is one of the main goals of the workshop. These participants commented that by continuously learning from the material scientists, the academia in computational mechanics can utilize the potential power of their computer codes to the maximum extent.

6.3 Theory, Methods and Tools for Additive Manufacturing Oriented Design and Optimization

6.3.1 Topic Discussion Organizers

Dr. Alonso Peralta (Honeywell Aerospace Inc., U.S.) and Prof. Xu Guo (Dalian University of Technology, China).

6.3.2 Summary

The topic discussion organizers, Dr. Alonso Peralta and Prof. Xu Guo, collected and posted the questions regarding to theory, methods and tools for additive manufacturing oriented design and optimization, including

- Short term goal: How to develop the topology optimization framework which is suitable to deal with AM related design and optimization problems considering, for example, overhang constraints, support costs, building orientation, residual stresses, anisotropies, robustness, fail-safe and uncertainties.
- How to consider the effect of manufacturing process on material properties for AM related topology optimization and design? How to incorporate this effect into optimization model and develop the corresponding solution algorithms?
- How to fully characterize and validate the failure mode of cellular/lattice materials and structures?
- How to design stretch-dominant micro-structured materials?
- What is the right methodology to incorporate both microstructure and topology in a load-bearing structure design?
- It is needed to have multiscale simulation methods for analysis and design of graded cellular/lattice structures considering more engineering oriented performances (e.g., strength, fracture, fatigue).

- It is critical to be able to design cellular/lattice structures with continuously varying/graded micro-structure cells, cell size, shape, and properties. To develop multiscale modelling/optimization methods for problems without strict separation of scales.
- A long term dream: High resolution real-time, full blown multiscale, multi-physics and multidisplinary modelling and design methods and software.
- How do you incorporate Reliability Engineering/Probabilistic design principles into Topology Optimization?
- What is the need for Design Space Exploration Techniques? One provides a set of rules and the computer does its magic.
- What are the limitation imposed by current optimization solution schemes and what kind of improvements are needed to overcome such limitations?
- For the AM structures, the shapes and the residual stresses can be predicted by the sequentially coupled thermo-mechanical model. But the controlling of the residual state is strongly dependent on the optimizations of the AM processing parameters. How to develop special optimization method for complex structures is always important for theoretical developments of AM and its relations to applications. The self-optimization design through controlling scanning paths and related parameters can possibly lead to nearly zero distortions and minimized residual stresses. The post heat treatment effect can be also included in this point. An integrated optimization of the processing parameters and the parameters for post heat treatments should be developed. How to select the important data can affect the data-driven design for AM.
- How to produce the best design to meet the performance requirements?

A participant from industry stated that AM is bottom-up, not top-down, giving people a lot of freedom. He does not think there is any issue related to IP because any one can do in his/her own way. The same participant is interested in physical understanding between micro and macro. He stated that one critical issue is what the most important is among so many variables and effects in AM.

6.4   Additive Manufacturing Experimental Methods and Results, and Additive Manufacturing Scalability

6.4.1   Topic Discussion Organizers

Prof. Lyle E. Levine (National Institute of Standards and Technology, U.S.), Prof. Anthony Rollett (Carnegie Mellon University, U.S.) and Prof. Pinlian Han (Southern University of Science and Technology, China).

6.4.2   Summary

The topic discussion organizers stated several key points in this topic including the determination of the measurement data, the measurement capabilities in validation of numerical models, qualification of AM parts, the flaw detection and eventual feedback systems, the establishments of benchmark tests, the standards for AM and the store and curate experimental data. They identified the following key points to be further investigated:

- Accurate modeling of AM processes and microstructure evolution requires parameters and functional behaviors that can only be obtained through measurements. What data are most needed and why?
- Large residual stresses from the AM build process occur at length scales ranging from the size of the part features down to the sub-grain structure. Emerging stress measurement techniques cover this full range. What are the critical questions these measurements could answer?
- Measurement experts need to know what information is needed before they develop new measurement methods. If it were possible to measure anything having to do with AM, what would you want measurement data for and why?
- Validation testing for AM parts poses unique challenges including, but not limited to, dimensional measurements of complex internal structures and flaw detection. What types of measurement capabilities are required for validation and what are the barriers to implementation?
- Qualification of AM parts requires consideration of the full life cycle behavior and failure within both operating and extreme conditions. What AM-unique issues need to be addressed?
- In situ process monitoring is critical for flaw detection and eventual feedback systems. What do we need to measure and how do we measure it?
- Surface roughness plays a critical role in fatigue behavior and cannot currently be removed from internal surfaces. What methods exist for effectively characterizing the surface roughness and surface cracks in AM-built materials?
- AM-Bench aims to produce rigorous, highly controlled additive manufacturing benchmark test data for validation testing of AM modeling methods and codes. What kinds of modeling predictions should AM-Bench start with and what kinds of benchmark tests would be needed?
- Nearly all AM roadmaps call for the development of standards. What kinds of standards are needed by the community?
- How should we store and curate the experimental data?

The participants discussed the key points in experimental measurements for additive manufacturing. Some of them mentioned the importance for the determination of fundamental information such as CCT and TTT diagrams especially for AM material. They discussed whether the AM data can be exactly repeated in experiments in the case of different power size and spacing for each experiment. They mentioned the experimental tests for validation purpose in different length scales. Some of them stated the useful information which needs to be addressed in experiments including the group energy densities, the roundness, and the porosity. Some of them suggested that industrial CT may be an available method for comparison purpose. Some of them suggested crystal plasticity in the modelling of fatigue with consideration of inclusions for comparison purpose with experimental fatigue tests. They mentioned the competitions of fatigue, corrosion and creep for the final failure modes of AM parts. Some of them suggested the development of new theories on fatigue description especially for AM. Some of them mentioned that the formation of standards for AM has been started in China under the guidance of Ministry of Industry and Information Technology in Chinese government.

Appendix A

A.1 Registered Workshop Participants

Alonso Peralta, Honeywell Aerospace Inc., U.S.
Anthony Rollett, Carnegie Mellon University, U.S.
Benben Ma, Dalian University of Technology, China
Bendong Xing, Shenyang Aircraft Design and Research Institute of AVIC, China
Bin Wu, Shenyang Aircraft Design and Research Institute of AVIC, China
Bin Ji, Shanghai Aerospace System Engineering Institute, China
Bingheng Lu, Xi'an Jiaotong University, China
Bo Wang, Dalian University of Technology, China
Chang Liu, Dalian University of Technology, China
Chenggang Jiang, Dalian University of Technology, China
Chunze Yan, Huazhong University of Science and Technology, China
Dehao Xu, Southern University of Science and Technology, China
Dongjiang Wu, Dalian University of Technology, China
Ernst Rank, Technische Universität München, Germany
Feng Lin, Tsinghua University, China
Gengdong Cheng, Dalian University of Technology, China
Hongfei Ye, Dalian University of Technology, China
Hongquan Hao, National Natural Science Foundation of China, China
Hualin Fan, Nanjing University of Aeronautics and Astronautics, China
Hui Li, Dalian University of Technology, China
Huiling Duan, Peking University, China
Jiachen Wang, Beijing Institute of Technology, China
Jiayong Zhang, Dalian University of Technology, China
Jimin Chen, Beijing University of Technology, China
Jin Wang, Dalian University of Technology, China
Jingyuan Wang, Dalian University of Technology, China
Jun Yan, Dalian University of Technology, China
Kathie Bailey, National Academy of Sciences, U.S.

© The Author(s) 2018
X. Guo et al., *Report of the Workshop Predictive Theoretical, Computational and Experimental Approaches for Additive Manufacturing (WAM 2016)*, SpringerBriefs in Applied Sciences and Technology, https://doi.org/10.1007/978-3-319-63670-2

Lars-Erik Lindgren, Luleå University of Technology, Sweden
Lei Wei, Northwestern Polytechnical University, China
Lianfeng Wang, Shanghai Aerospace Equipment Manufacture, China
Lyle E. Levine, National Institute of Standards and Technology, U.S.
Michele Chiumenti, Technical University of Catalonia, Spain
Pengying Yang, Peking University, China
Pinlian Han, Southern University of Science and Technology, China
Qi Zhang, Nanjing University of Aeronautics and Astronautics, China
Qinglin Duan, Dalian University of Technology, China
Qingxi Hu, Shanghai University, China
Quhao Li, Dalian University of Technology, China
Riye Xue, Dalian University of Technology, China
Rui Li, Dalian University of Technology, China
Shan Tang, Chongqing University, China
Shasha Qiu, Dalian University of Technology, China
Shutian Liu, Dalian University of Technology, China
Songtao Chen, Dalian University of Technology, China
Tianjian Lu, Xi'an Jiaotong University, China
Ting Wen, Xi'an Jiaotong University, China
Michael Yu Wang, The Hong Kong University of Science and Technology, China
Wei Yang, National Natural Science Foundation of China, China
Weibin Wen, Beijing Institute of Technology, China
Weidong Huang, Northwestern Polytechnical University, China
Weisheng Zhang, Dalian University of Technology, China
Wenzheng Wu, Jilin University, China
Wing-Kam Liu, Northwestern University, U.S.
Wuqing He, Southern University of Science and Technology, China
Xiangming Wang, Shenyang Aircraft Design and Research Institute of AVIC, China
Xu Guo, Dalian University of Technology, China
Xuyang Liu, Chongqing University, China
Ya Qian, Tsinghua University, China
Yiqian He, Dalian University of Technology, China
Yonggang Zheng, Dalian University of Technology, China
Yongtao Lv, Dalian University of Technology, China
Yongtao Wang, Dalian University of Technology, China
Zhao Zhang, Dalian University of Technology, China
Zhen Chen, University of Missouri, U.S.
Zhi Sun, Dalian University of Technology, China
Zhirui Fan, Dalian University of Technology, China
Zongliang Du, Dalian University of Technology, China

Appendix B

B.1 Workshop Agenda

DAY 1: OCTOBER 17, 2016

8:00 a.m. Introduction

Dongming Guo, Dalian University of Technology (President of Dalian University of Technology)

Wei Yang, National Natural Science Foundation of China (Chair of the National Natural Science Foundation of China and President of the Chinese Society of Theoretical and Applied Mechanics)

Gengdong Cheng, Dalian University of Technology (Chair of the Workshop)

Wing-Kam Liu, Northwestern University (Co-Chair of the Workshop, Director of the International Research Center for Computational Mechanics of Dalian University of Technology)

8:40 a.m. Session 1: Overview and Theoretical Studies

Facilitator	Xu Guo (Dalian University of Technology, China)
Speakers	Kathie Bailey (National Academies of Sciences, U.S.),
	Bingheng Lu (Xi'an Jiaotong University, China),
	Wing-Kam Liu (Northwestern University, U.S.)

10:40 a.m. Session 2: Simulation and Modelling

Facilitator	Michele Chiumenti (Technical University of Catalonia, Spain),
	Feng Lin (Tsinghua University, China)
Speakers	Ernst Rank (Technische Universität München, Germany),
	Weidong Huang (Northwestern Polytechnical University, China)

1:30 p.m. Session 3: Simulation and Modelling

Facilitator	Michael Yu Wang (Hong Kong University of Science and Technology, China),
	Zhao Zhang (Dalian University of Technology, China)

© The Author(s) 2018
X. Guo et al., *Report of the Workshop Predictive Theoretical, Computational and Experimental Approaches for Additive Manufacturing (WAM 2016)*, SpringerBriefs in Applied Sciences and Technology, https://doi.org/10.1007/978-3-319-63670-2

Speakers   Feng Lin (Tsinghua University, China),
            Chunze Yan (Huazhong University of Science and Technology,
            China),
            Alonso Peralta (Honeywell Aerospace Inc., U.S.)

3:50 p.m. **Session 4: Simulation and Modelling**

Facilitator Jun Yan (Dalian University of Technology, China),
            Yonggang Zheng (Dalian University of Technology, China)
Speakers   Huiling Duan (Peking University, China),
            Michele Chiumenti (Technical University of Catalonia, Spain),
            Lars-Erik Lindgren (Luleå University of Technology, Sweden),
            Zhao Zhang (Dalian University of Technology, China)

6:30 p.m. **Adjourn Day 1**
DAY 2: OCTOBER 18, 2016
8:00 a.m. **Session 5: Design**

Facilitator Pinlian Han (Southern University of Science and Technology, China),
            Qinglin Duan (Dalian University of Technology, China)
Speakers   Ole Sigmund (Technical University of Denmark, Denmark),
            Xu Guo and Shutian Liu (Dalian University of Technology, China),
            Michael Yu Wang (Hong Kong University of Science and
            Technology, China)

10:20 a.m. **Session 6: Design**

Facilitator Huiling Duan (Peking University, China),
            Jimin Chen (Beijing University of Technology, China)
Speakers   Pinlian Han (Southern University of Science and Technology, China),
            Xiangming Wang (Shenyang Aircraft Design and Research Institute
            of AVIC, China)

1:30 p.m. **Session 7: Experimental Studies and Others**

Facilitator Lars-Erik Lindgren (Luleå University of Technology, Sweden),
            Xu Guo (Dalian University of Technology, China)
Speakers   Lyle E. Levine (National Institute of Standards and Technology, U.S.),
            Anthony Rollett (Carnegie Mellon University, U.S.),
            Dongjiang Wu (Dalian University of Technology, China)

3:50 p.m. **Session 8: Experimental Studies and Others**

Facilitator   Lyle E. Levine (National Institute of Standards and Technology, U.S.),
               Zhen Chen (University of Missouri, U.S.)

Speakers   Lianfeng Wang (Shanghai Aerospace Equipment Manufacture,
               China),
               Jimin Chen (Beijing University of Technology, China),
               Yonggang Zheng (Dalian University of Technology, China)

5:50 p.m. **Adjourn Day 2**
DAY 3: OCTOBER 19, 2016
8:30 a.m. **Session 9: Discussion**

Facilitator   Wing-Kam Liu (Northwestern University, U.S.),
               Gengdong Cheng (Dalian University of Technology, China)

11:30 a.m. **Adjourn Workshop**

References

N. Aage, E. Andreassen, B.S. Lazarov, O. Sigmund, Giga-resolution computational morphogenesis—Aircraft wings and bird beaks. Submitted (2016)

J. Alexandersen, O. Sigmund, N. Aage, Large scale three-dimensional topology optimization of heat sinks cooled by natural convection. Int. J. Heat Mass Transf. **100**, 876–891 (2016)

J. Alexandersen, N. Aage, C.S. Andreasen, O. Sigmund, Topology optimization for natural convection problems. Int. J. Numer. Meth. Fluids **76**(10), 699–721 (2014)

H. Bikas, P. Stavropoulos, G. Chryssolouris, Additive manufacturing methods and modelling approaches: a critical review. Int. J. Adv. Manuf. Technol. **83**, 389–405 (2016)

D. Butnaru, Computational steering with reduced complexity. PhD Thesis, Technical University of Munich (2013)

M. Chiumenti, M. Cervera, N. Dialami, B. Wu, L. Jinwei, C. Agelet de Saracibar, Numerical modeling of the electron beam welding and its experimental validation. Finite Elem. Anal. Des. **121**, 118–133 (2016)

M. Chiumenti, X. Lin, M. Cervera, W. Lei, Y. Zheng, W.D. Huang, Numerical simulation and experimental calibration of additive manufacturing by blown powder technology. Part I: thermal analysis. Rapid Prototyp. J. **23**, 448–463 (2017)

M. Chiumenti, M. Cervera, A. Salmi, C.A. de Saracibar, N. Dialami, K. Matsui, Finite element modeling of multi-pass welding and shaped metal deposition processes. Comput. Methods Appl. Mech. Eng. **199**(37), 2343–2359 (2010)

M. Chiumenti, E. Neiva, E. Salsi, M. Cervera, S. Badia, C. Davies, and C. Lee, Numerical modelling and experimental validation in selective laser melting. Addit. Manuf. Submitted (2017)

A. Clausen, N. Aage, O. Sigmund, Topology optimization of interface problems and coated structures. Comput. Methods Appl. Mech. Eng. **290**, 524–541 (2015)

A. Clausen, N. Aage, O. Sigmund, Exploiting additive manufacturing infill in topology optimization for improved buckling load. Eng. **2**(2), 250–257 (2016)

A. Clausen, F. Wang, J.S. Jensen, O. Sigmund, J.A. Lewis, Topology optimized architectures with programmable Poisson's ratio over large deformations. Adv. Mater. **27**(37), 5523–5527 (2015)

J.P. Groen, O. Sigmund, Homogenization-based topology optimization for high-resolution manufacturable micro-structures. Submitted (2016)

R. Cunningham, S.P. Narra, T. Ozturk, J. Beuth, A.D. Rollett, Evaluating the effect of processing parameters on porosity in electron beam melted Ti-6Al-4V via synchrotron X-ray microtomography. JOM **68**(3), 765–771 (2016)

J.W. Elmer, J. Vaja, H.D. Carlton, R. Pong, The effect of Ar and N_2 shielding gas on laser weld porosity in steel, stainless steels, and nickel. Weld. Res. **94**(10), 313s–325s (2015)

M. Fisk, J. Andersson, R. du Rietz, S. Haas, S. Hall, Precipitate evolution in the early stages of ageing in Inconel 718 investigated using small-angle x-ray scattering. Mater. Sci. Eng. A **612**, 202–207 (2014)

© The Author(s) 2018

X. Guo et al., *Report of the Workshop Predictive Theoretical, Computational and Experimental Approaches for Additive Manufacturing (WAM 2016)*, SpringerBriefs in Applied Sciences and Technology, https://doi.org/10.1007/978-3-319-63670-2

M. Fisk, A. Lundbäck, Simulation and validation of repair welding and heat treatment of an alloy 718 plate. Finite Elem. Anal. Des. **58**, 66–73 (2012)

W.D. Huang, X. Lin, Research progress in laser solid forming of high-performance metallic components at the state key laboratory of solidification processing of China. 3D Print. Addit. Manuf. **1**, 156–165 (2014)

Y. Idell, L.E. Levine, A.J. Allen, F. Zhang, C.E. Campbell, G.B. Olson, J. Gong, D.R. Snyder, H. Z. Deutchman, Unexpected beta-phase formation in additive-manufactured Ni-based superalloy. JOM **68**(3), 950–959 (2016)

G. Jonathan, A. Karim, 3D printing in pharmaceutics: A new tool for designing customized drug delivery systems. Int. J. Pharm. **499**(1–2), 376–394 (2016)

S. Kollmannsberger, A. Özcan, M. Carraturo, N. Zander, E. Rank, Computational modeling of selective laser melting: accurate resolution of moving phase change effects via a dynamic hp-discretization. www.cie.bgu.tum.de. Submitted (2016)

L.E. Levine, C. Okoro, R. Xu, Full elastic strain and stress tensor measurements from individual dislocation cells in copper through-Si vias. IUCrJ **2**(6), 635–642 (2015)

P. Li, D.H. Warner, A. Fatemi, N. Phan, Critical assessment of the fatigue performance of additively manufactured Ti–6Al–4 V and perspective for future research. Int. J. Fatigue **85**, 130–143 (2016)

L.E. Lindgren, A. Lundbäck, M. Fisk, R. Pederson, J. Andersson, Simulation of additive manufacturing using coupled constitutive and microstructure models. Addit. Manuf. In press (2016)

L. Liu, J. Yan, G.D. Cheng, Optimum structure with homogeneous optimum truss-like material. Comput. Struct. **86**(13–14), 1417–1425 (2008)

M.B. Liu, G.R. Liu, *Particle-based Methods for Multi-scale and Multi-physics*, World Scientific (2016)

A. Lundbäck, L.E. Lindgren, Modelling of metal deposition. Finite Elem. Anal. Des. **47**, 1169–1177 (2011)

P. Lv, Y.H. Xue, Y. Shi, H. Lin, H.L. Duan, Metastable states and wetting transition of submerged superhydrophobic structures. Phys. Rev. Lett. **112**, 196101 (2014)

J. Madamesila, P. McGeachy, J.E.V. Barajas, R. Khan, Characterizing 3D printing in the fabrication of variable density phantoms for quality assurance of radiotherapy. Physica Med. **32**(1), 242–247 (2016)

V. Manvatkar, A. De, T. DebRoy, Heat transfer and material flow during laser assisted multi-layer additive manufacturing. J. Appl. Phys. **116**(12), 124905 (2014)

P. Michaleris, Modeling metal deposition in heat transfer analyses of additive manufacturing processes. Finite Elem. Anal. Des. **86**, 51–60 (2014)

J.C. Michel, H. Moulinec, P. Suquet, Effective properties of composite materials with periodic microstructure: a computational approach. Comput. Methods Appl. Mech. Eng. **172**(1), 109–143 (1999)

O.D. Neikov, I.B. Murashova, N.A. Yefimov, S. Naboychenko, *Handbook of non-ferrous metal powders: technologies and applications*, Elsevier (2009)

F.Y. Niu, D.J. Wu, S. Yan, G.Y. Ma, B. Zhang, Process optimization for suppressing cracks in laser engineered net shaping of Al_2O_3 ceramics. *SFF Symp* Aug 8 ∼ 11, Austin, Texas, U.S. (2016)

F.Y. Niu, D.J. Wu, S.Y. Zhou, G.Y. Ma, Power prediction for laser engineered net shaping of Al_2O_3 creamic parts. J. Eur. Ceram. Soc. **34**(15), 3811–3817 (2014)

S.C. O'Keeffe, S. Tang, A.M. Kopacz, J. Smith, D.J. Rowenhorst, G. Spanos, W.K. Liu, G.B. Olson, Multiscale ductile fracture integrating tomographic characterization and 3D simulation. Acta Mater. **82**, 503–510 (2015)

H.K. Rafi, N.V. Karthik, H. Gong, T.L. Starr, B.E. Stucker, Microstructures and mechanical properties of Ti6Al4 V parts fabricated by selective laser melting and electron beam melting. J. Mater. Eng. Perform. **22**(12), 3872–3883 (2013)

S. Rao, R. Cunningham, T. Ozturk, A.D. Rollett, Measurement and analysis of porosity in Al-10Si-1 Mg components additively manufactured by selective laser melting. Mater. Perform. Characterization **5**(5). doi:10.1520/MPC20160037 (2016)

D. Rosenthal, Mathematical theory of heat distribution during welding and cutting. Weld. J. **20**(5), 220s–234s (1941)

J.C. Schuren, P.A. Shade, J.V. Bernier, S.F. Li, B. Blank, J. Lind, P. Kenesei, U. Lienert, R.M. Suter, T.J. Turner, D.M. Dimiduk, J. Almer, New opportunities for quantitative tracking of polycrystal responses in three dimensions. Curr. Opin. Solid State Mater. Sci. **19**(4), 235–244 (2015)

O. Sigmund, On the usefulness of non-gradient approaches in topology optimization. Structural and Multidisciplinary Optimization **43**(5), 589–596 (2011)

O. Sigmund, K. Maute, Topology optimization approaches. Structural and Multidisciplinary Optimization **48**(6), 1031–1055 (2013)

O. Sigmund, N. Aage, E. Andreassen, On the (non-) optimality of Michell structures. Structural and Multidisciplinary Optimization **54**, 361–373 (2016)

J. Smith, W. Xiong, J. Cao, W.K. Liu, Thermodynamically consistent microstructure prediction of additively manufactured materials. Comput. Mech. **57**(3), 359–370 (2016)

J. Smith, W. Xiong, W. Yan, S. Lin, P. Cheng, O.L. Kafka, G.J. Wagner, J. Cao, W.K. Liu, Linking process, structure, property, and performance for metal-based additive manufacturing: computational approaches with experimental support. Comput. Mech. **57**(4), 583–610 (2016)

M. Tang, P.C. Pistorius, Oxides, porosity and fatigue performance of AlSi10 Mg parts produced by selective laser melting. Int. J. Fatigue **94**(2), 192–201 (2017)

J. Tao, Y.G. Zheng, Z. Chen, H.W. Zhang, Generalized interpolation material point method for coupled thermo-mechanical processes. Int. J. Mech. Mater. Des. **12**, 577–595 (2016)

F. Wang, B.S. Lazarov, O. Sigmund, On projection methods, convergence and robust formulations in topology optimization. Structural and Multidisciplinary Optimization **43**(6), 767–784 (2011)

J. Wu, N. Aage, R. Westermann, O. Sigmund, Infill Optimization for Additive Manufacturing-Approaching Bone-like Porous Structures. To Appear Transactions on Visualization and Computer Graphics, arXiv preprint arXiv **1608**, 04366 (2017)

Y.Q. Wang, M.Y. Wang, F.F. Chen, Structure-material integrated design by level sets. Structural and Multidisciplinary Optimization (2016). doi:10.1007/s00158-016-1430-5

Y.Q. Wang, F.F. Chen, M.Y. Wang, Concurrent design with connectable graded microstructures. Comput. Methods Appl. Mech. Eng. **317**, 84–101 (2017)

Y.Q. Wang, M.Y. Wang, F.F. Chen, Multiscale structural design with graded microstructures via a mathematical optimization for additive manufacturing. Under preparation

Z. Wang, Q. Zhang, K. Zhang, G. Hu, Tunable digital metamaterial for broadband vibration isolation at low frequency. Adv. Mater. (2016). doi:10.1002/adma.201604009

L. Wei, X. Lin, M. Wang, W.D. Huang, Cellular automaton simulation of the molten pool of laser solid forming process. Acta Phys. Sinica **64**, 018103 (2015)

X. Wei, L. Mao, R.A. Soler-Crespo, J.T. Paci, J. Huang, S.T. Nguyen, H.D. Espinosa, Plasticity and ductility in graphene oxide through a mechanochemically induced damage tolerance mechanism. Nat. Commun. **6**, 8029 (2015)

Y.H. Xue, Y. Yang, H. Sun, X.Y. Li, S. Wu, A.Y. Cao, H.L. Duan, A switchable and compressible carbon nanotube sponge electrocapillary imbiber. Adv. Mater. **27**, 7241–7246 (2015)

C. Yan, L. Hao, A. Hussein, P. Young, J. Huang, W. Zhu, Microstructure and mechanical properties of aluminium alloy cellular lattice structures manufactured by direct metal laser sintering. Mater. Sci. Eng. A **628**, 238–246 (2015)

C. Yan, L. Hao, A. Hussein, P. Young, Ti-6Al-4 V triply periodic minimal surface structures for bone implants fabricated via selective laser melting. J. Mech. Behav. Biomed. Mater. **51C**, 61–73 (2015)

W. Yan, W. Ge, J. Smith, S. Lin, O.L. Kafka, F. Lin, W.K. Liu, Multi-scale modeling of electron beam melting of functionally graded materials. Acta Mater. **115**, 403–412 (2016)

W. Yan, J. Smith, W. Ge, F. Lin, W.K. Liu, Multiscale modeling of electron beam and substrate interaction: a new heat source model. Comput. Mech. **56**(2), 265–276 (2015)

N. Zander, T. Bog, S. Kollmannsberger, D. Schillinger, E. Rank, Multi-level hp-adaptivity: high-order mesh adaptivity without the difficulties of constraining hanging nodes. Comput. Mech. **55**(3), 499–517 (2015)

F. Zhang, L.E. Levine, A.J. Allen, C.E. Campbell, A.A. Creuziger, N. Kazantseva, J. Ilavsky, In situ structural characterization of ageing kinetics in aluminum alloy 2024 across angstrom-to-micrometer length scales. Acta Mater. **111**, 385–398 (2016)

Y.G. Zheng, C. Chen, H.F. Ye, H.W. Zhang, Lattice Boltzmann models for the grain growth in polycrystalline systems. AIP Adv. **6**, 085315 (2016)

Printed in the United States
By Bookmasters